523 · 8446

17. JUL 1982

13. MAR. 1986

2 0 FEB 1982

1 0 FEB 1988

1 0 FEB 1982

1 OCT 1986

- 4 MAR 1982

25. 3. 82   2 5 NOV 1986

2 2 SEP 1982

1 5 SEP 1982

28 DEC 1985

METROPOLITAN
BRADFORD   LIBRARIES
This book must be returned by the last date
entered above. An extension of loan may
be arranged if the book is not in demand.

SUPERNOVAE
AND THEIR REMNANTS

SUPERNOVAE AND THEIR REMNANTS

Edited by

Peter J. Brancazio

*Brooklyn College, the City University of New York and
Belfer Graduate School of Science, Yeshiva University*

and

A.G.W. Cameron

*Belfer Graduate School of Science, Yeshiva University and
Institute for Space Studies, Goddard Space Flight Center, NASA*

Proceedings of the Conference on Supernovae, held at the
Goddard Institute for Space Studies, NASA 1967

GORDON AND BREACH SCIENCE PUBLISHERS

New York   London   Paris

Copyright © 1969 by GORDON AND BREACH, SCIENCE PUBLISHERS, INC.
150 Fifth Avenue, New York, N. Y. 10011

Library of Congress catalog card number: 78–84/92

Editorial office for the United Kingdom:
Gordon and Breach, Science Publishers Ltd.
12 Bloomsbury Way
London W.C.1

Editorial office for France:
Gordon & Breach
7–9 rue Emile Dubois
Paris 14ᵉ

Distributed in Canada by:
The Ryerson Press
299 Queen Street West
Toronto 2B, Ontario

METROPOLITAN
BRADFORD LIBRARIES
5 1081 Cᵤ
5 23.
8446   21.1.82
4 586526 01

All rights reserved. No part of this book may be reproduced or utilized in any form or by any means, electronic or mechanical, including photocopying, recording, or by any information storage and retrieval system, without permission in writing from the publishers. Printed in east Germany

PREFACE

On November 2–3, 1967, the Institute for Space Studies of the Goddard Space Flight Center, National Aeronatics and Space Administration, was host to an international group of astronomers and physicists at a conference on supernovae. This was one of a continuing series of interdisciplinary meetings on topics in space physics held at the Institute. The conference was organized by A. G W. Cameron and was jointly sponsored by the Belfer Graduate School of Science of Yeshiva University and by the Institute.

Nova explosions have been a familiar phenomenon in astronomy for a very long time; records of "guest stars" go back in Chinese and other chronicles for more than 2000 years. However, it was only in the 1930's that it was realized that a small minority of the nova explosions represented stellar outbursts orders of magnitude more brilliant than the others, so that a separate classification became desirable. It is now recognized that a supernova explosion produces a major disruption of a star, whereas a nova explosion is a relatively minor phenomenon which blows off only 10^{-4} or 10^{-5} of the mass. Two of the contributions at this conference are from astronomers who were prominent in the early investigations of supernovae and who have maintained a long interest in them: F. Zwicky and R. Minkowski. They report on the current observational situation.

The paper by A. Poveda and L. Woltjer reports on two major developments by these authors, the demonstration of a relationship between supernova remnants and x-ray sources, and interpretations of supernova remnant observations which indicate that the energy released in the supernova explosions was significantly less than is typically found in the supernova hydrodynamic calculations. The latter point caused much controversy at the conference. The papers by R. Schwartz, S. Colgate, and D. Arnett are expositions and defenses of the hydrodynamic calculations. The participants at the conference started to refer to the "observational supernova" and to the "theoretical supernova" as two separate entities which might or might not have properties in common.

The papers by J. Truran, A. Cameron, K. Thorne, and L. Sartori and P. Morrison deal with various aspects of the supernova explosion which may or may not tell us much about the explosion in terms of observable features. These aspects include the nuclear constitution of the ejected matter, the properties of supernova remants, assumed to be neutron stars, and the shape of the supernova light curve.

The final set of papers, by P. Morrison, J. Scargle, and D. Melrose, deals with additional properties of ejected supernova envelopes.

Since the conference C. Hansen and D. Arnett have found possible stellar evolutionary pathways which may lead to the purely nuclear explosion of a star. In some respects such explosions may resemble the "observational supernova" more than is the case in the hydrodynamic calculations which lead to the implosive formation of a neutron star remnant. In some respects this is helpful from the nucleosynthesis point of view, since the "theoretical supernova" appears to lead to much too large a production of heavy neutron-rich nuclides. However, the existence of these nuclides implies in any case that at least a small minority of the supernovae must involve the full hydrodynamic implosion.

Shortly after the conference the interesting new phenomenon of pulsars was discovered. One possible explanation of pulsars which seems to suffer a minimum of theoretical difficulties compared to other proposals is radio flashes from a rotating neutron star. If this should prove to be correct, we will have an additional tool to investigate supernova remnants.

The conference was tape-recorded. These tape records formed the basis for the preparation of first drafts of the papers. These were sent to the authors, who then had the opportunity to revise and update their papers as they wished. We thank them for their cooperation. We have special thanks for Mrs. Enid Silva, who served as conference secretary, handled very capably the arrangements for the meeting, and participated extensively in the preparation of this proceedings volume. We thank Dr. Robert Jastrow and Dr. Arthur Levine, Director and Executive Officer of the Institute, for their cooperation in assuring the success of the conference. Mr. George Goodstadt, Mr. Nicholas Panagakos, Mr. Larry Stein, and several other staff members of the Institute gave valuable assistance in the operation of the conference. Mrs. Ronnie Brancazio gave valuable assistance in the preparation of the proceedings.

P. J. Brancazio
A. G. W. Cameron

CONTENTS

1

SOME RESULTS OF THE INTERNATIONAL SEARCH FOR SUPERNOVAE

F. Zwicky

Mount Wilson and Palomar Observatories
Carnegie Institution of Washington
California Institute of Technology

A   On the Search and Discovery of Supernovae

1   Outline of topics to be discussed

IN DISCUSSING the status of the search for supernovae I would like to describe some of the methods used in the past and the results achieved. It will also be useful to consider the most desirable projects to be carried out in the future and possible new methods of search. Some of the characteristics of various types of supernovae will be discussed as well as the statistics pertaining to the frequency of different types of supernovae. Finally a few suggestions will be made about various types of phenomena that may possibly be related to supernova events.

2   On the initiation and the pursuit of the search for supernovae

The first extragalactic supernova was discovered by Hartwig at Dorpat (Muller and Hartwig, 1920) on Aug. 31, 1885 near to the nucleus of the great nebula in Andromeda (Messier 31). By the 1920's about a dozen supernovae had been found in what are now known to be extragalactic stellar systems. But the nature and significance of these objects was not realized at that time and, in contradistinction to the advent of common novae, no search for them was conducted.

In 1932 a group was formed consisting of Baade, Humason, Minkowski and myself for the purpose of searching for and investigating supernovae. Between 1933 and 1936 I surveyed the Virgo Cluster and parts of the Ursa Major Cloud with a 3.5 inch Wollensack camera mounted on an experimental 12-inch reflector, which we had installed on the roof of the astrophysics building at the California Institute of Technology in Pasadena and which I used for guiding the camera. As ill luck would have it, no bright enough supernova flared up during the three years in the two clusters mentioned. Supernovae of type I in these clusters would have had an apparent photographic magnitude $m_p \sim 11.0$ at maximum, while my camera reached $m_p = 13.5$ on clear nights. Some of the fainter objects of course might have been missed.

Because of the above-mentioned failure, I would have abandoned the search if the opportunity had not offered itself in 1935 to build a powerful Schmidt telescope. After a brief visit to Schmidt at the Bergedorf Observatory in 1934 we were able, with the enthusiastic support of the construction group of the 200-inch telescope (Dr. John A. Anderson, Mr. Russell W. Porter and others) to convince Dr. George Ellery Hale to let us build an 18″ Schmidt camera to be used for the supernova search and to serve generally as a scout telescope for the 200-inch telescope. The 18-inch Schmidt was built in 12 months and put in operation on the night of September 5, 1936. In parenthesis it may be mentioned that its construction has proved to be one of the most fruitful projects of all times. In addition to the supernovae, between 1936 and 1941 forty-eight Humason-Zwicky stars were discovered; the first dwarf galaxies [Sextans, Leo I etc.] (Zwicky 1942a), and intergalactic bridges between widely separated galaxies were found, and many new clusters of galaxies were detected showing that most galaxies are members of clusters rather than being field galaxies, as it had been erroneously inferred by Hubble (1936).

The allocation of tasks then was as follows: the author and one part-time assistant, Dr. J. J. Johnson, would survey about 4000 galaxies as often as possible. Baade would follow up the light curves, and Minkowski and Humason would investigate the spectra using the large reflectors of the Mount Wilson Observatory. Fortunately we had the full support of Dr. W. S. Adams, Director of the Mount Wilson Observatory, for our projects. Whenever a supernova was discovered the observers at the large reflectors had to yield their nights to one of the investigators of supernovae. In recent years, unfortunately, we have not had the same type of cooperation and much potential information on discovered supernovae has thus been lost.

3 Number of supernovae discovered

From September 1936 until the end of 1941 my assistant and I conducted an intensive search with the 18-inch Schmidt covering about 10,000 square degrees; that is, one-fourth of the sky. In any given month, of course, only about one half to two-thirds of this area could be searched, the remainder passing by in the daytime. About 3000 galaxies brighter than the apparent photographic magnitude $m_p = 15.0$ and about 700 galaxies brighter than $m_p = 13.0$ were searched whenever the season and the weather permitted it. Special attention was paid the nearby galaxies such as Messier 31, 32, 33, 51, 81, 82, 100, IC 10, 1613, NGC 2403, 6822, 6946, etc.

The method of searching used involved the comparison of films by super-position under a binocular microscope. Successive old films were used for this comparison, so that any supernova missed originally would have been found later on. We can thus be fairly certain that no supernova brighter than about $m_L - 0.5$ was missed where $m_L \sim 17.3$ is the average limiting magnitude for stellar images on the films used.

Various other methods of comparison were tried but eventually abandoned as too uncertain or too time-consuming. For instance, recent (negative) films were superposed on older positive films and scanned or projected. Defects in the emulsions, impossibility of exact registry and dust particles rendered this method unusable, especially if photoelectric scanning was attempted instead of visual inspection.

A special blink apparatus was also built, which is of excellent quality, but whose use proved to be too time-consuming because of the difficulty of properly mounting the films. In the first four years I discovered 14 "live" supernovae and one old one in NGC 3184 on plates obtained with the 100-inch telescope on Mount Wilson in 1921.

My assistant Dr. J. J. Johnson, a meticulous worker, was content to take very beautiful pictures, but showed no interest to search them for super-novae. Only at the beginning of World War II when I became busy with other problems did he start to examine the films, and he discovered four supernovae in 1940 and 1941.

As far as we know all of the 14 supernovae that I found were of Type I (except possibly but most uncertainly SN 1937a in NGC 4157). On the other hand Johnson's first two objects in NGC 5907 and NGC 4725 proved to be beautiful examples of what we designated as of Type II. The characteristics of the various types will be discussed later.

Actually 35 years had to pass during which I discovered 37 supernovae until I found my first bona fide supernova (SN 1967h) of Type II in NGC 4254. Minkowski's law, that Zwicky discovers only supernovae of Type I and Johnson only objects of Type II, was finally shown to allow of some exceptions.

During World War II the search was essentially stopped. It was only resumed seriously late in the 1950's after the 48-inch Schmidt became available and after successively more observatories offered cooperation both for the search and the follow-up investigations. At present the following observatories are participating:

a) Survey of the brighter galaxies (Shapley-Ames and NGC-type objects) with medium-size Schmidt cameras or wide angle refractors at Burakan, Abastumani au Nauchny in the USSR, Konkoly in Hungary, Tautenburg (East Germany), Asiago (Italy), Meudon (France), Zimmerwald (Switzerland), Pretoria (South Africa), Córdoba (Argentina), Tonantzintla (Mexico), and, in part, Stromlo (Australia).

b) Survey of two dozen clusters of galaxies with the 48-inch Schmidt at Palomar (in addition to the use of the 18-inch Schmidt for the brighter galaxies).

c) Survey of the nearest galaxies with the 8-inch F/1 Schmidt of the California Institute of Technology which I installed on the grounds of the Lick Observatory with the efficient cooperation of Mr. Eugene Harlan.

The numbers of supernovae so far discovered in extra-galactic systems since 1885 are listed in Table I.

Table 1   Rate of discovery of supernovae

Year	Cumulative total of known supernovae	Year	Cumulative total of known supernovae
1885	1	1950	49
1900	3	1955	53
1910	7	1960	82
1920	13	1964	152
1930	19	1967	204
1940	38		

The number corresponding to a given year gives the cumulative total of known supernovae as of that year.

The discovery rate during the past seven years is thus on the average 17 supernovae per year. An additional seven supernovae were found in January, February and March 1968, so that the total since 1885 now stands at 211. Of these about 160 were discovered by the author and the searchers who have worked with him directly at Palomar (Johnson, Wild, Humason, Gates, Gomes, Mendez, Rudnicki, Berger, Kearns, Reeves and Kowal).

4 Possible and suggested methods of search

To all of those who have been engaged in the search for supernovae it has become painfully clear that it has been a gruelling enterprise, and all have therefore looked for more efficient methods.

Unfortunately nothing, as far as I am concerned, has as yet been suggested which would be effective as far as the discovery of supernovae and at the same time would insure the *preservation of the immense wealth of additional information* which is stored and easily accessible on the photographic plates and films which we have taken during the many years of our surveys.

For instance, in 1939 my friend Dr. V. K. Zworykin, director of research for the Radio Corporation of America, became interested in Schmidt cameras for use in television. In this connection Dr. Zworykin and I often considered the possibility of sweeping the sky with a few Schmidt telescopes installed in a few well placed locations and televising the information to say a thousand of the ten thousands of amateur astronomers all over the world. Each amateur would have been instructed to watch a limited area, comparing it with a standard view of this same area, and thus singling out supernovae, novae, variable stars, flare stars, and the like. Discoveries could have been immediately transmitted telephonically or by wire to various observatories. This beautiful scheme proved impracticable since we knew of no way to really organize the amateurs effectively. But even if we had succeeded in doing so we could not have reproduced the information storage now available on our photographic plates.

As a second possibility we also considered tracking fields of say 50 to 100 square degrees one after another and to scan electronically each field in succession, feeding the information into a computer and comparing it with previously stored records. This method would have avoided the necessity of slewing the telescopes with mechanically intolerable speeds from one galaxy to another and overall corrections for the sky glow and thin obscuring clouds could have been made. Simultaneous photographic storage of the actual distribution and the light intensity from the whole field of celestial objects could also have been made. This then would have been far more effective than some of the proposed methods of scanning individual galaxies one after another for the appearance of supernovae, without securing the permanent storage of all information on the ever changing aspects of the celestial fields.

In the course of my surveys of the skies during three and a half decades I have come to the conviction, however, that the most effective method for the future will be one which will allow us to discover supernovae faster and at the same time store all of the information retained by the photographic plates. In addition, however, we must ask for more. I refer in particular to information on the spectra of new objects found right at the moment of discovery. This goal obviously can readily be achieved through the use of full size objective transmission gratings mounted on Schmidt camera or equivalent wide angle telescopes.

The tremendous potential value of a large Schmidt telescope equipped with a full-size transmission echelette grating of proper ruling is obvious (Zwicky 1967a). Such a grating would, for instance, produce one first-order spectrum with over 90% of all of the light entering the telescope in it. Two to three percent might be in the direct image and the rest in the remaining orders of the spectrum. The dispersion is essentially linear, thus avoiding crowding in any parts of the spectrum. The presence of direct images permits the direct evaluation of the apparent magnitudes of all objects registered, as well as the determination of the redshifts. For supernovae, common novae, flare stars, eruptions of compact galaxies, and quasars, both magnitudes and spectra are available on the discovery plates. With the 48-inch Palomar Schmidt one might preferably use gratings with 150 to 200 lines/inch or 375 to 500 lines/mm producing dispersions from 555 to 417 Å/mm. At such dispersions, and using black and white emulsions (for instance Eastman IIa-O), ordinary stars could be classified to the 15th apparent magnitude and various types of supernovae, quasars and emission line compact galaxies be discovered at between the 15th and 17th apparent photographic magnitude.

If such an objective grating had been available for the 48-inch Schmidt telescope, quasars and compact emission line galaxies would have been discovered fifteen years earlier, and the discovery and the gathering of information on supernovae would have been vastly accelerated. The feasibility of the production of large objective echelette transmission gratings and their usefulness had actually been demonstrated by the author as early as 1939

(Zwicky 1941, 1962). In 1941 he and Professor R. W. Wood also built three full size mosaic objective echelette gratings for the 18-inch Schmidt telescope with which many emission line stars and novae, such as Nova Cygni of 1942, were discovered (Zwicky 1942b). Unfortunately our efforts did not strike much resonance among our colleagues and no further large echelette transmission gratings have been built. As far as the discovery of supernovae is concerned our old idea of using fast electronic scanning and evaluation of the records by computer is now being put into practice by some agencies. The promoters of these projects, however, seem to overlook that little will be gained by upping the discovery rate. Indeed, as far as the statistics of frequency of occurrence of supernovae is concerned we would need thousands of objects in order to improve materially on our presently available results, so that the cost involved in the projects mentioned is very high indeed. It is of interest that the cost per object of the 160 supernovae that were discovered by members of my group has been about $1000 on the average, including the construction of the 8-inch and 18-inch Schmidt telescope and housings, auxiliary instruments, films, plates, salaries etc., but not the construction of the 48-inch Schmidt.

In view of our experiences I therefore propose to promote the construction of several echelette transmission gratings for Schmidt telescopes of apertures greater than one meter to be installed in both the northern and southern hemispheres. In addition to all other uses mentioned, plates obtained with such equipment will serve the search for supernovae in the following ways:

a) Supernovae can be recognized on single plates from their spectra. The necessity of comparing (blinking) two plates is eliminated.

b) Both spectrum and direct image of the supernova are available at discovery.

c) Intergalactic supernovae will be discovered. These have so far escaped us, and they will also escape electronic survey methods which scan only the galaxies. The author had at one time attempted to search the full fields of direct images but the occurrence of too many variable stars defeated his efforts.

d) Telescopes equipped with objective gratings could of course be used most advantageously in combination with electronic scanning and computers, producing the information referred to in the previous sections. The storage of all other information which is so valuable when photographic plates are available, however, would be lost.

5   Frequency of supernovae

In the diagrams of Fig. 1 the distribution of the 211 supernovae discovered in the periods from 1885 to 1930, 1930 to 1960, and 1960 to March 31, 1968 is plotted.

An example of a beautiful supernova which I discovered belatedly in 1968 on Palomar Sky Survey plates of 1955 in NGC 4335 is shown in Fig. 2. A negative print (Fig. 2a) without the supernova and a positive print (Fig. 2b) showing the supernova have been chosen in order to demonstrate the quality of reproduction of the various objects.

A first attempt to derive the frequency of supernovae was made by Zwicky (1938) two years after the beginning of the systematic search with the 18-inch Schmidt telescope on Mt. Palomar. A frequency of one supernova per 460 years per average galaxy of the Shapley-Ames Catalogue was obtained. This value was later revised (Zwicky 1942c) on the basis of the search with the same instrument in the period from September 1936 to January 1, 1940. Five supernovae had been found in 837 galaxies of the Shapley-Ames Catalogue. Taking into account the limiting magnitudes of the photographic records available and the so-called control times, a frequency

$$\nu = \text{one supernovae per 359 years per Shapley-Ames galaxy}$$

was obtained. No additional accurate determination of ν was made until recently, when it was found that the frequency of supernovae among the member galaxies in the center of the Coma Cluster which are brighter than $m_p = 15.7*$ is also of the order of one supernova per galaxy per 300 years (Zwicky 1965a, 1965b).

In November 1967, when the count stood at 203 supernovae since 1885, Dr. R. Barbon surveyed the pertinent frequencies for a homogeneous sample of CATALOG galaxies, for galaxies in some of the large clusters and for a selection of stellar systems in the Local Group.

Using all the data available, Barbon essentially confirms my original result; namely, that the frequency of occurrence of supernovae in galaxies whose apparent photographic luminosities lie within three to four magnitudes of the brightest normal stellar systems is of the order of one on the average in 300 years per galaxy, regardless of whether they are field galaxies or cluster galaxies.

---

* $m_p = 15.7$ is the apparent photographic magnitude of galaxies chosen for inclusion in our six-volume *Catalogue of Galaxies and Clusters of Galaxies* (Zwicky *et al.* 1961–1968) called in the following the CATALOG.

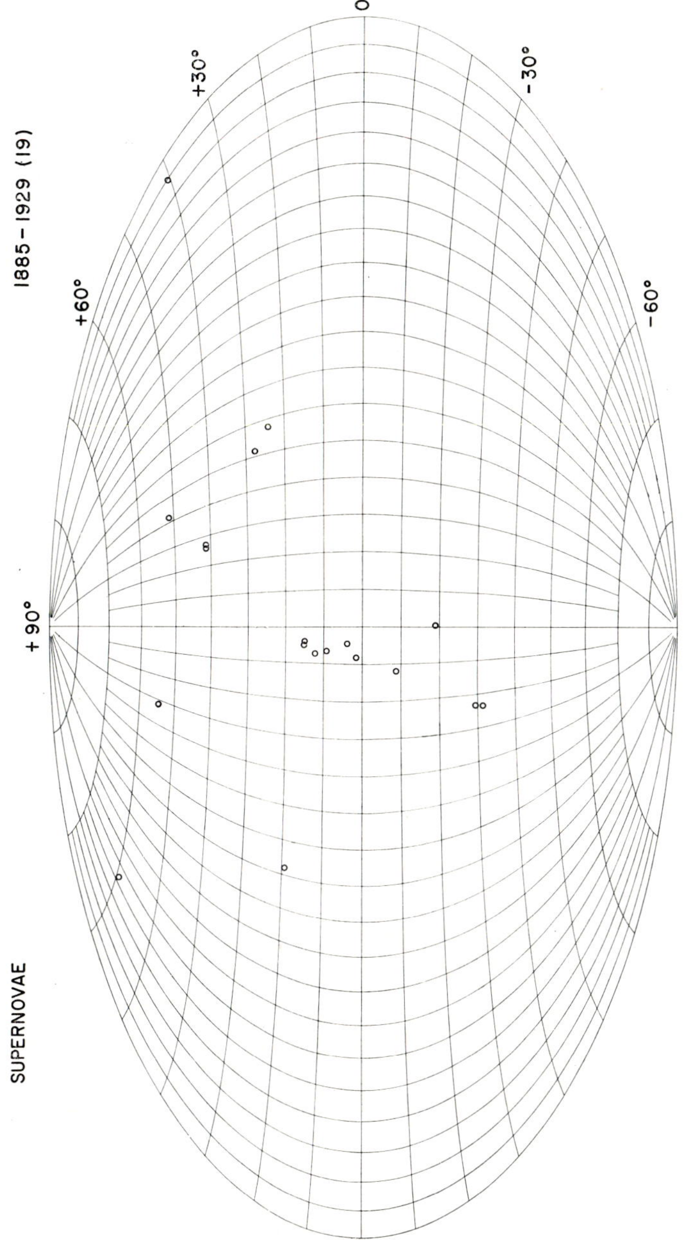

Fig. 1a Distribution over the sky of supernovae that flared up in the period from 1885 to the end of 1929.

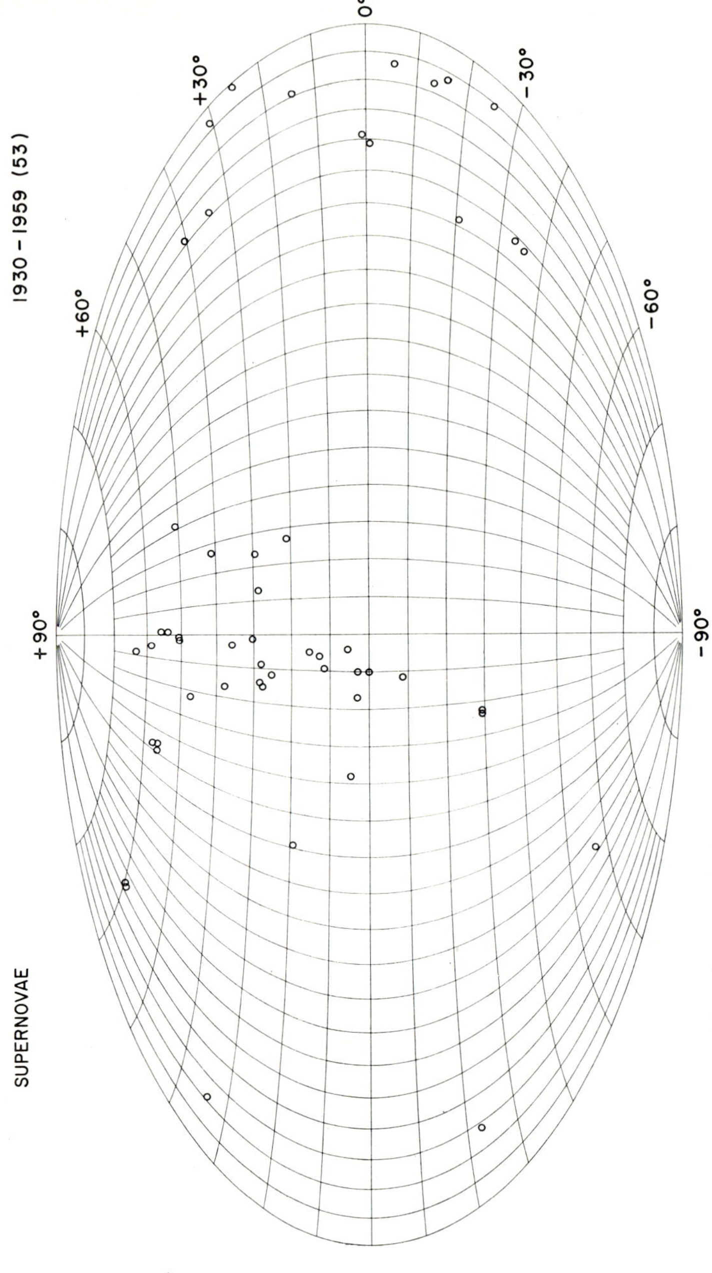

Fig. 1b Distribution of supernovae that occurred in the period from 1930 to 1959.

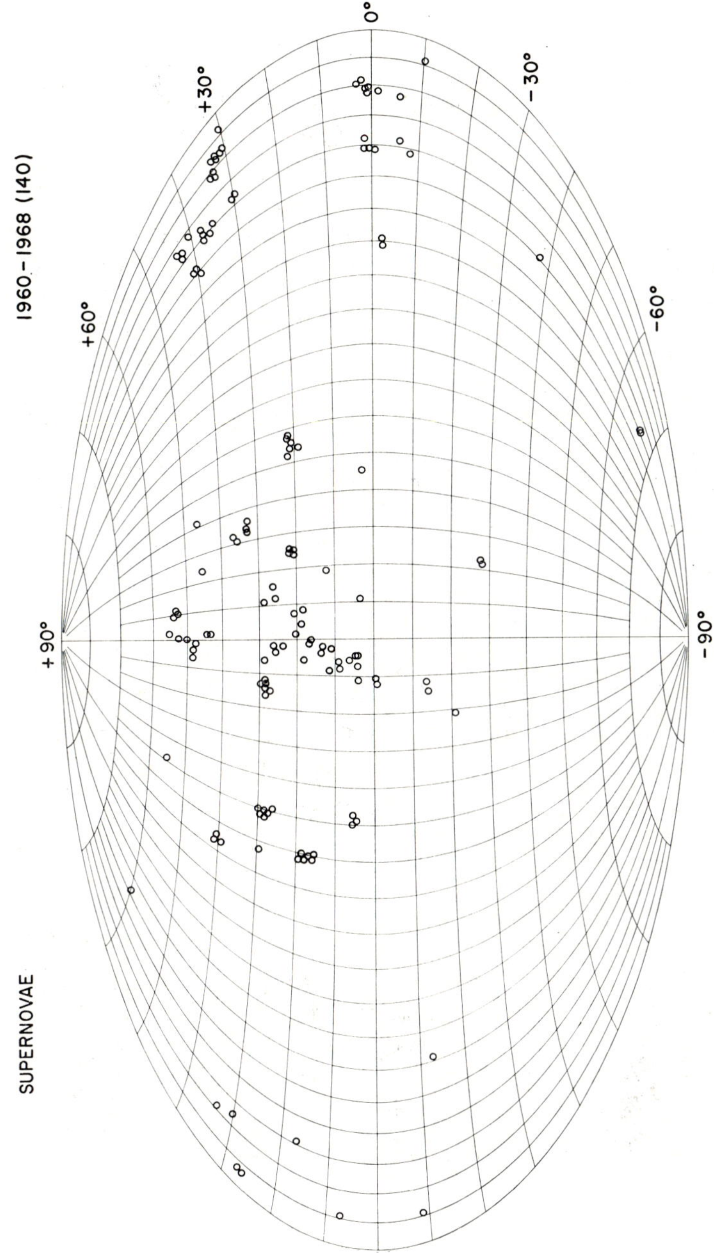

Fig. 1c Supernovae in the period from 1960 until March 31, 1968.

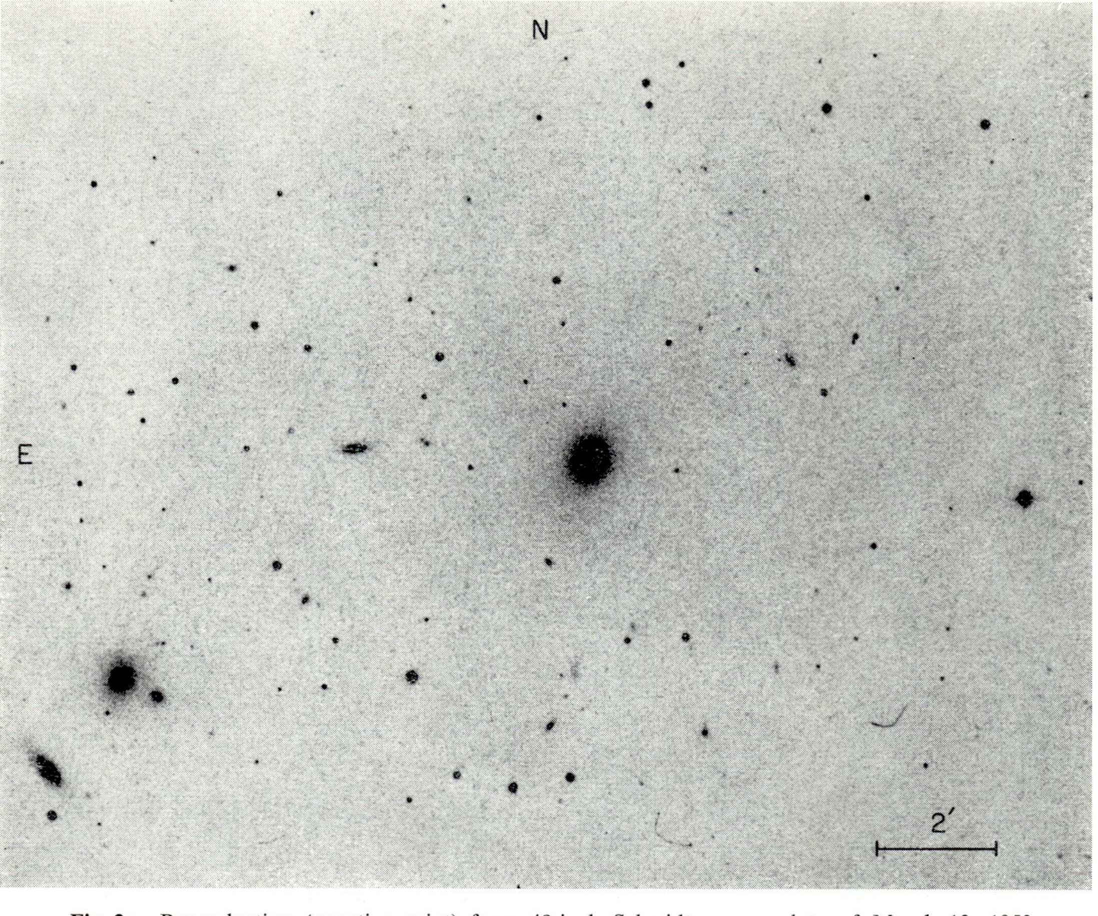

Fig. 2a Reproduction (negative print) from 48-inch Schmidt survey plate of March 13, 1953, showing the elliptical galaxy NGC 4335 at R.A. $12^h20^m.6$ and Decl. $+58°44'$ (1950), $m_p = 13.7$ in the center, without supernova.

Fig. 2b   Reproduction (positive print) from 48-inch Schmidt survey plate of May 10, 1955 showing the supernova SN 1555e at apparent photographic magnitude $m_p = 16.3$, 19″ West and 7″ South of the center of NGC 4335.

A number of investigators during the past twenty years have tried to demonstrate that the frequency in question is as high as 1 supernova per 50 years per bright galaxy, or even greater (for instance Katgert and Oort, 1967). If this were the case the searchers at the 14 observatories who are now participating in the surveys organized by the Committee of the I.A.U. for research on supernovae would better quit the business since it would mean that they have missed seven out of eight supernovae on their plates. All we can say is that those who have not searched themselves, but who disagree with us so violently, are quite welcome to examine the ten thousands of plates and films we have in our possession and prove through the discovery of the expected one thousand additional supernovae, that the present searchers have been essentially blind.

It lies beyond the scope of the present condensed report to analyze in detail where the statistical considerations made by the enthusiastic adherents of greater abundances of supernovae are incorrect. We only mention the characteristics of some of the basic errors committed.

a) The investigators in question have never themselves searched for supernovae or acquainted themselves with the intricate aspects of the search through thorough consultation with any of the old hands among the searchers.

b) Some of the investigators in question have committed the fatal error of choosing their sample of galaxies after the fact, so to speak, and not through *a priori* selection totally independent of any of the discoveries achieved. Through such improper and biased selection almost any result can be achieved. For instance, one might arbitrarily choose a group of the nearest well known galaxies, such as NGC 147, 185, 205, 221, 224, 300, 404, 598, 6822, and IC 10, 342, 1613, 4182 plus SMC and LMC. In these 15 galaxies one supernova has been discovered in the last 40 years, that is an average of one supernova per 600 years per galaxy. The frequency would become still smaller if we insisted to take into account the fact that in NGC 205, 221, 224, 598 probably no supernova was missed in the last 80 years and almost certainly in SMC and LMC in the last two to three hundred years.

If, on the other hand we should insist on including NGC 6946 in our local group above, while omitting many other possible candidates, we should have 16 galaxies in which between 1927 and 1967 three supernovae were found, resulting in a frequency of one supernova per galaxy in 213 years, that is three times the previously derived value.

c) None of the enthusiasts of abundance of supernovae has investigated the statistics of large samples. The necessity for very large samples is clearly apparent when we consider the case of three clusters. The Cancer cluster has about 100 members brighter than $m_p = 15.7$. During our search at Mt. Palomar we found six supernovae in these galaxies, giving a frequency of about one supernova per galaxy per 100 years on the average.

The field of the Hercules cluster covered by us has about 150 galaxies brighter than $m_p = 15.7$. These are structurally of the same varied character as those in the Cancer cluster. They produced so far only one bona fide supernova, making the frequency about one per galaxy per 1000 years; that is, *ten times smaller* than for the Cancer cluster.

The frequency of supernovae in the brighter galaxies of the Coma cluster, on the other hand, is of the order of one per galaxy per 400 years, close to the average found for all galaxies.

Most interesting for the statistics of supernovae and possibly for our future understanding of the material contents of galaxies are those stellar systems which seem to produce supernovae at about ten times the average rate. This suggests the following considerations.

Among about 1000 Shapley-Ames galaxies (Shapley and Ames, 1932), supernovae have appeared in 65 of them since 1885, that is about one supernova per fifteen galaxies. We therefore should expect 65/15 double events, that is four galaxies which have produced 2 supernovae each from 1885 to the end of 1967. As it happens this expectation is exactly fulfilled inasmuch as two supernovae have appeared in NGC 2841 (SN 1912a, SN 1957a), NGC 3631 (SN 1964a, SN 1965l), NGC 3938 (SN 1961u, SN 1964l) and in NGC 4157 (SN 1937a, SN 1955a).

For triple events the number of supernova parent galaxies expected among one thousand Shapley-Ames Galaxies is $65/15^2 = 65/225 < 1/3$. That is, practically no Shapley-Ames Galaxy should be expected to have produced three or more supernovae in the period from 1885 to the end of 1967. In reality three supernovae have flared up in each of the following galaxies, NGC 3184 (SN 1921b, SN 1921c, SN 1937f), NGC 4303 (SN 1926a, SN 1961i, SN 1964f); NGC 4321 (SN 1901b, 1914a, 1959e); NGC 5236 (SN 1923a, SN 1957d, SN 1950b) and in NGC 6946 (SN 1917a, SN 1939c, SN 1948b, SN 1968d). This demonstrates clearly that there exist galaxies, so far all spirals of probably only medium indicative luminosity, that produce ten times as many supernovae as other structurally very similar stellar systems. The assumption by Katgert and Oort (1967) and others that all galaxies of the same type and luminosity produce supernovae at a rate of one in fifty years is therefore erroneous.

Based on the above statistics I predicted during the New York Symposium that we should expect the above-mentioned galaxies *to continue producing supernovae at a high rate*. Indeed after my return to Pasadena, Wild from Zimmerwald (Berne, Switzerland) wired that he had just discovered a *fourth* supernova in NGC 6946.

Attention should finally be called to the following circumstances. First, I have shown previously (Zwicky 1965a) that we certainly have missed some supernovae which have appeared within the luminous nuclei, cores and central disks of spherical and elliptical galaxies, as well as So, Sa, Sb, SA and SB systems. I estimate that for this reason the overall frequency will be found to be about twenty percent greater than Barbon and I originally estimated (Zwicky 1965b). Secondly a search must now be started for the appearance of supernovae in compact galaxies. Some of these have been found to change brightness abruptly, for instance, in the compact galaxy at R.A. $0^h39^m33^s$, Decl. $+40°03'$ (1950) (Zwicky 1965c); some of these changes are almost certainly due to the flare-up of supernovae or chains of supernovae.

Finally, a question of great interest is what the frequency of various types of supernovae in various types of galaxies might be. To answer this, however, it will almost certainly be necessary to discover and investigate in detail a few thousand supernovae, a goal which does not appear readily attainable in the near future.

Generally speaking, the statistics of the search for supernovae has been hexed in many ways. For instance, the weather during the four months from July to Nov. 1967 has been excellent both in Europe and on this continent. At all of the observatories involved more pictures have been taken during these months than over periods of years before that, and no one has found any supernoval. On the other hand, I found two within an hour in June 1967 and three on three successive days in March 1965. My former collaborator at Palomar, Mr. Paul Wild, has the habit of shooting pictures through a single clear hole in an otherwise completely covered sky and discovering supernovae on films which cover only a handful of galaxies. Yet he has had no luck whatever surveying rich fields like those containing the Virgo cluster of galaxies.

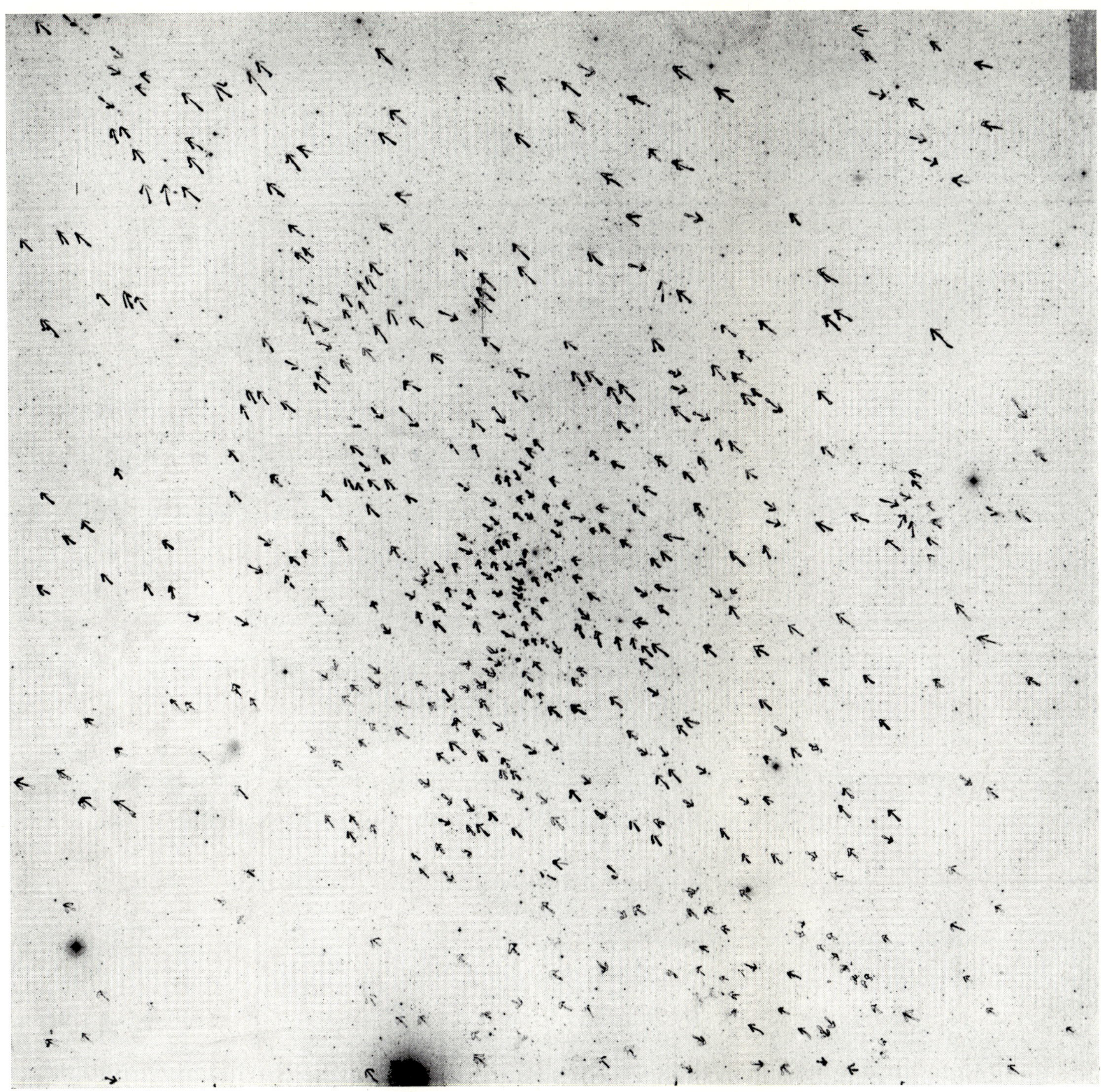

Fig. 3 Reproduction of a $10' \times 10'$ photograph with the 48-inch Schmidt centered on the Perseus cluster. Arrows mark the many galaxies which Dr. M. L. Humason examined one by one with a hand lens in his gruelling search for supernovae in 1959 before a blink apparatus became available. Two dozen fields were surveyed in this laborious fashion. (The author's hat is off to this indefatigable observer.)

B Types of Supernovae

I would now like to discuss some of the problems involved in identifying the various types of supernovae. We find that it is not yet possible to clearly distinguish the type by means of light curves or colors. It is therefore necessary to classify them essentially only by their spectra. However, for a Type I supernova, if whatever is ejected does not interact with gas clouds or produce secondary emission, the light curve will have a fairly definite shape (Zwicky 1965a). The slope of the declining portion of the light curve can be from 1 magnitude in 40 days to 1 magnitude in 100 days. We have also noticed that all of the Type I supernovae in the Coma Cluster are of magnitude 16 ± 0.5. So it appears that a Type I supernova is a relatively well-defined object.

In identifying supernova types from their spectra, I rely on the blue part of the spectrum. For a Type I, there are generally five bands in the blue and a tremendous depression, reduced to zero red shift, at $\lambda_d = 3790$ Å on the average. Then the λ/λ_d (which is independent of the red shift) for the five bands are 1.280, 1.210, 1.132, 1.048, and 0.976. The results are remarkably uniform for all except one of the 60 Type I supernovae. In that particular case there was a sixth band at $\lambda/\lambda_d = 0.946$. We do not yet know how to interpret any of the bands.

Type II supernovae do not show as much uniformity. Although a great many spectra have been obtained of Type II, the spectra are not fully understood. Usually Hα, Hβ, Hγ and Hδ stand out clearly, but the other bands have not been interpreted with certainty. The widths of these bands indicate a velocity of expansion of about 6000 km/sec.

Spectra of Type III supernovae are the same as Type II except that the bands are twice as wide. In NGC 4303, the bands were shifted towards the violet for the first six weeks, indicating an expanding shell of gas that couldn't be seen through. Within a week the bands were shifted the full width to the red and one could see the receding parts of the shell as well. The light curves for a Type III do not drop off nearly as rapidly as those for Type II supernovae. All of these observations indicate that considerably more mass is thrown out by a Type III.

We have only two or three of Type IV, in which we see only two bands instead of five in the blue, and these bands shift very rapidly in intensity (Zwicky 1965a).

We know of Type V supernovae*, in NGC 1058 and 3631; both were discovered by Wild within a year. Their light curves are similar to those of ordinary novae, but they must be supernovae since they are seen in quite distant galaxies. Furthermore, in an ordinary nova forbidden lines will appear within a month, whereas the mentioned two Type V supernovae have not yet shown any forbidden lines after four years. Mostly the Balmer lines and Fe II lines are seen in emission. This again indicates that much mass was thrown out in the explosion. The expansion velocity for the supernova in NGC 1058 is about 2100 km/sec.

In the past, whenever we didn't have any good spectra we classified supernovae on the basis of their ultraviolet output. We had previously found that Type I supernovae have remarkably little intensity in the ultraviolet, so that we automatically classified a supernova as Type II if it were strong in the ultraviolet when no spectrum was available. However, Chalonge has recently observed what seems to be a Type II supernova in NGC 3389 that shows practically no ultraviolet. Consequently, the question of the ultraviolet output of supernovae will have to be considered thoroughly.

It would be of great value to us if we could see the remnant of an object which had previously been observed telescopically as a supernova. There is a possibility that this can be achieved for the supernova in IC 4182, which was first seen in August 1937. It now still seems to be visible at about the 23rd magnitude in the 200-inch telescope. The use of image tubes will greatly facilitate the study of this object.

With regard to the distribution of supernovae, there seems to be no particular distinction as to the type of galaxy in which they appear. The distribution is roughly proportional to the relative numbers of normal spirals, barred spirals, elliptical galaxies and irregular galaxies. An important quantity worth mentioning is Δ, the angular distance from the center of the galaxy to the supernova. If we plot (Zwicky 1965a) the number of supernovae seen against $V_s \cdot \Delta$, where V_s is the symbolic velocity of the system, we can see that we have apparently lost supernovae in the centers of galaxies. As I mentioned above we are now attempting to remedy that situation.

Another useful statistic is related to the magnitude of the galaxy minus the magnitude of the supernova at maximum. The range observed so far extends from a supernova which was 5 magnitudes brighter than its parent galaxy to one which was 7 magnitudes dimmer than its parent galaxy. The

---

* η Carinae of 1843 in our galaxy is also thought to be a supernova of Type V.

frequency distribution of this quantity gives a curve which allows us to derive a luminosity for galaxies which agrees with the assumption that supernovae occur in numbers roughly proportional to the luminous content of galaxies (Zwicky 1965a).

I would like to mention that we are now preparing a complete list, along with the relevant data, of all supernovae seen since 1885. We hope to have this information available by January 1969.

C On the Theory of Supernovae

Three tasks remain for the theory of supernovae. First, the facts available must be properly interpreted. Second, the basic causes for the various types of supernovae and the phenomena associated with them must be established. And third, plans must be made for future significant projects of observation and theory.

1 Interpretations of the facts available

As mentioned before, supernovae of five types have so far been found as identified from the characteristics of their spectra. While Types II, III and V show identifiable emission lines and bands, the spectra of Type I and IV have not been interpreted in spite of continued attempts by spectroscopists during more than 30 years to do so. The task of correctly identifying any or all features of the spectra of supernovae of Type I thus appears to be *the most difficult problem* which has so far presented itself in the whole long history of spectroscopy.

2 Basic causes for supernova outbursts

Many theoretical models have been proposed to explain the occurrence of supernova outbursts. But none of these models has so far been definitely linked to any particular type of supernova. I only mention here three of the possibilities which I have suggested myself during the past 30 years. These are

(i) The collapse of stars becoming unstable into neutron stars (Baade and Zwicky 1934a, 1934b; Zwicky 1939).

(ii) The collapse of large quiescent gas clouds becoming unstable into neutron stars (Zwicky 1957).

(iii) The collision of stars at high speeds and the resulting formation of neutron stars (Zwicky 1966, 1967 b). This process is of importance only in very compact galaxies, compact parts of galaxies and quasars.

Neutron stars are only being "precipitated" out of the collisions of stars that have relative velocities of thousands of kilometers/second. The welding together of stars at low velocities and subsequent formation of neutron stars proposed by Colgate (1967) cannot really take place because of the impossibility of conserving angular momentum.

3   Planning of future projects

The following important developments associated with supernova explosions remain to be investigated.

(i) The emergence of forbidden lines at a stage I when the expanding gas clouds have become sufficiently tenuous.

(ii) The emission of radio waves at a stage II when tenuous plasmas and extended magnetic fields have come into existence.

(iii) A stage III at which X-rays are emitted.

(iv) A stage IV at which cosmic rays are emitted.

At the present time we have no knowledge of the times which elapse between the implosion and outburst and the stages I, II, III, IV.

As to the most desirable instrumentation for the near future I refer again to section A 4 above and to the bibliography given below.

References

Baade, W., and Zwicky, F. 1934a, *Proc. Nat. Acad. Sciences*, **20**, 254.
– 1934b, ibid, **20**, 259.
Colgate, S.A. 1967, *Ap. J.*, **150**, 163.
Hubble, E.P. 1936, *The Realm of the Nebulae* (New Haven, Conn.: Yale Univ. press).
Katgert, P., and Oort, J.H. 1967, *B.A.N.*, **19**, 239.
Müller, G., and Hartwig, E. 1920, *Geschichte und Literatur des Lichtwechsels* II, p.417
   (Leipzig: Poeschel & Trepte).
Shapley, H., and Ames, A. 1932, *Survey of the External Galaxies Brighter than the 13th Magnitude* (Cambridge, Mass.: Harvard Observatory).
Zwicky, F. 1938, *Ap. J.*, **88**, 529.
 –1939, *Phys. Rev.*, **55**, 726.

Zwicky, F. 1941, *P.A.S.P.*, **53**, 242.
- 1942a, *Phys. Rev.*, **61**, 489.
- 1942b, *P.A.S.P.*, **54**, 206.
- 1942c, *Ap. J.*, **96**, 28.
- 1957, *Morphological Astronomy* (Berlin: Springer Verlag).
- 1962, *Morphology of Propulsive Power* (Monograph No. 1 of the Society for Morphological Research) (Pasadena, Calif.: The Society).
- 1965a, Chap. on Supernovae, in "Stellar Structure" ed. by L.H.Aller and D.B.McLaughlin, Vol. VIII of *Stars and Stellar Systems* (Chicago: Univ. of Chicago Press).
- 1965b, *I.A.U. Transactions*, Vol. XIIA, p. 407.
- 1965c, *Carnegie Yearbook 64*, in "Reports … Mount Wilson and Palomar Observatories", p. 33 (Wash. D.C.: Carnegie Inst. of Washington).
- 1966, *Proc. International Conference on Cosmology, Padova*, 1964 (Bologna: G.Barbera).
- 1967a, *Die Sterne*, **43**, 89.
- 1967b, *P.A.S.P.*, **79**, 443.
Zwicky, F., Herzog, E., Wild, P., Karpowicz, M., and Kowal, C. 1960–1968, *Catalogue of Galaxies and Clusters of Galaxies*, Vols. 1–6 (Pasadena: Calif. Inst. of Technology).

DISCUSSION

R. Minkowski There has recently been some discussion as to the frequency of supernovae. What you have done is to calculate the frequency only for the brightest galaxies. It is reasonable that a more luminous galaxy ought to have more supernovae, since it has more stars. In that connection, I would like to ask whether the galaxies which had 3 supernovae are particularly luminous. As far as I remember, they are not.

F. Zwicky That is correct, except for NGC 4321, where we are not certain whether to put it in front or in back of the Virgo cluster, so there is an uncertainty of 3 to 4 magnitudes.

P. Morrison What are the total numbers of events in the Coma, Cancer, and Hercules clusters?

F. Zwicky Twelve in Coma, five or six in Cancer, and one in Hercules during the last ten years.

A. Poveda I remember a claim that Sc galaxies and irregular galaxies had a higher frequency. Is that correct?

F. Zwicky Not that I can see.

R. Minkowski It is a question of the statistics. I think there are two supernovae in irregulars, and that is the basis for the statement.

A. Finzi Is it true that all of the galaxies in which three supernovae were found are Sc galaxies?

F. Zwicky That is true, but there are so many Sc galaxies in the Shapley-Ames catalog and relatively few ellipticals. However, it may be significant.

R. Minkowski Do you now have a Type II in an elliptical galaxy?

F. Zwicky I am not certain. The difficulty is that when galaxies are relatively small, one cannot say whether they are elliptical.

2

SUPERNOVAE
AND SUPERNOVA REMNANTS

R. Minkowski

Radio Astronomy Laboratory
University of California
Berkeley, California

DURING THE last few years information has become available which makes it possible and necessary to revise some of our ideas about supernova remnants and consequently about supernovae themselves (Minkowski 1967). One major source of this new information is Zwicky's supernova search. In particular, it is now clear that there are some very rare varieties of supernovae (which have been classified by Zwicky as Types III, IV, and V) about which we did not know originally. A second source of new information is provided by radio observations of the brightness distribution of supernova remnants; these measurements generally provide a much better picture of a supernova remnant than do optical observations.

Let us begin by summarizing the basic information on supernovae. Type I supernovae have a mean absolute visual magnitude at maximum of -19.0 ± 0.3 with a dispersion of 0.7 magnitudes. At maximum the visual magnitude is roughly equal to the photographic magnitude. After an initial drop of 2 to 3 magnitudes in 20 to 30 days, the light curve shows the well-known approximately exponential decay. The integrated observed luminosity, which is closely similar for all Type I supernovae, is about 3.6×10^{49} ergs during the outburst. We cannot determine the bolometric luminosity; in particular if there should be a large amount of x-ray emission, the total emitted radiation could be considerably larger. Type II supernovae are a much more individualistic group; the mean absolute visual magnitude is -17.7 ± 0.3 with a dispersion of 0.8 magnitudes. The scatter and the accuracy of this value are probably affected severely by observational selection, since there have been very few Type II supernovae for which the maximum has been observed. Their light curves are individually different, although there are some similarities. The curves tend in general to show some kind of a shoulder after the maximum, followed by a rather rapid decline. The height of the shoulder is about 1.5 magnitudes below the maximum. The initial decline is somewhat smaller than the initial decline in supernovae of Type I. The integrated luminosities differ widely over a range from 10^{48} to 10^{49} ergs/sec. This is again the observed luminosity; we cannot determine the bolometric luminosity. As far as we know Type I supernovae are Population II objects. They are observed in elliptical galaxies. It may have something to do with the great uniformity of the phenomenon that the brightest stars in Population II have a narrow mass range. These are the stars which will evolve most rapidly into supernovae. Type II supernovae are Population I objects; it is quite possible that the wide range of characteristics has to do with the fact that in Population I the most massive stars have a wide mass range of up to 20 or 30 solar masses. Two other types are of interest to us here: Type III supernovae are similar to Type II with an indication that Type III have a much higher mass; supernovae of Type V have the characteristics of an irregular variable, but reach supernova luminosities.

The spectra of Type I supernovae have not yet been satisfactorily interpreted. In the beginning there are very broad bands which narrow later on. Up to now it was not clear how the great widths of these lines were to be interpreted: do they mean very high velocities or a high degree of compositeness? In supernovae of Type II we have velocities of expansion of the order of 6000 km/sec with clear indications that ejections at different velocities occur; i.e., there are multiple absorptions as in ordinary novae. Type III supernovae start out with a high velocity of the order of 12,000 km/sec, which afterwards decreases to approximately 6000 km/sec. The only well-known example of Type V, the supernova in NGC 1058 (1961), has a velocity of expansion of about 2100 km/sec.

Now let us look at some of the results of the new information which we have acquired. We now have rather good observations of what is undoubtedly the remnant of a Type I supernova; namely, the remnant of Tycho's supernova of 1572. Partly as a consequence of these new observations, I believe that we can now state quite definitely that the Crab Nebula is not a remnant of a supernova of Type I. The reason why the Crab Nebula was thought to be a remnant of a Type I supernova is a rather curious one. The belief somehow arose naturally without much thinking because at the time when the

Crab Nebula was ascribed to the event observed by the Chinese in 1054, there was only one type of supernova known and that was Type I. As a result, the Crab Nebula automatically became a remnant of Type I, and no one ever questioned this afterwards. Moreover, people had so thoroughly convinced themselves of this assignment that doubt was cast on the clear evidence that Tycho's and Kepler's novae were actually supernovae of Type I, as evidenced by their light curves. Tycho and Kepler did not have photoelectric photometers, of course; their photometric observations were in fact comparisons with bright stars and planets. When a set of photometric observations is made, it must later be reduced to magnitudes. It doesn't matter whether this is done the following morning or 200 years later, and in the case of Tycho's and Kepler's supernovae a few hundred years passed before Baade reduced these light curves to magnitudes. There's nothing illegitimate about this, but under the impression that the Crab Nebula is a Type I remnant, people somehow got the idea that these light curves were artfully manufactured. This they were not. Figure 1 shows visual light curves for various types of supernovae. Also shown are the data of Tycho and Kepler reduced to visual magnitudes. The scatter is indicative of the understandably low accuracy of their observations, but there is really no doubt that the supernovae observed by Tycho and Kepler were of Type I.

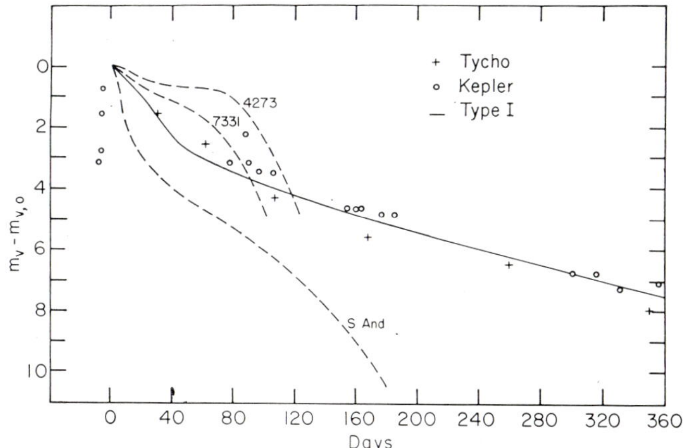

Fig. 1   Visual light curves of several supernovae. The solid line represents an average Type I supernova. The dashed lines give the light curves for several Type II supernovae. The crosses and circles represent the data for Tycho and Kepler, as reduced to magnitudes by Baade.

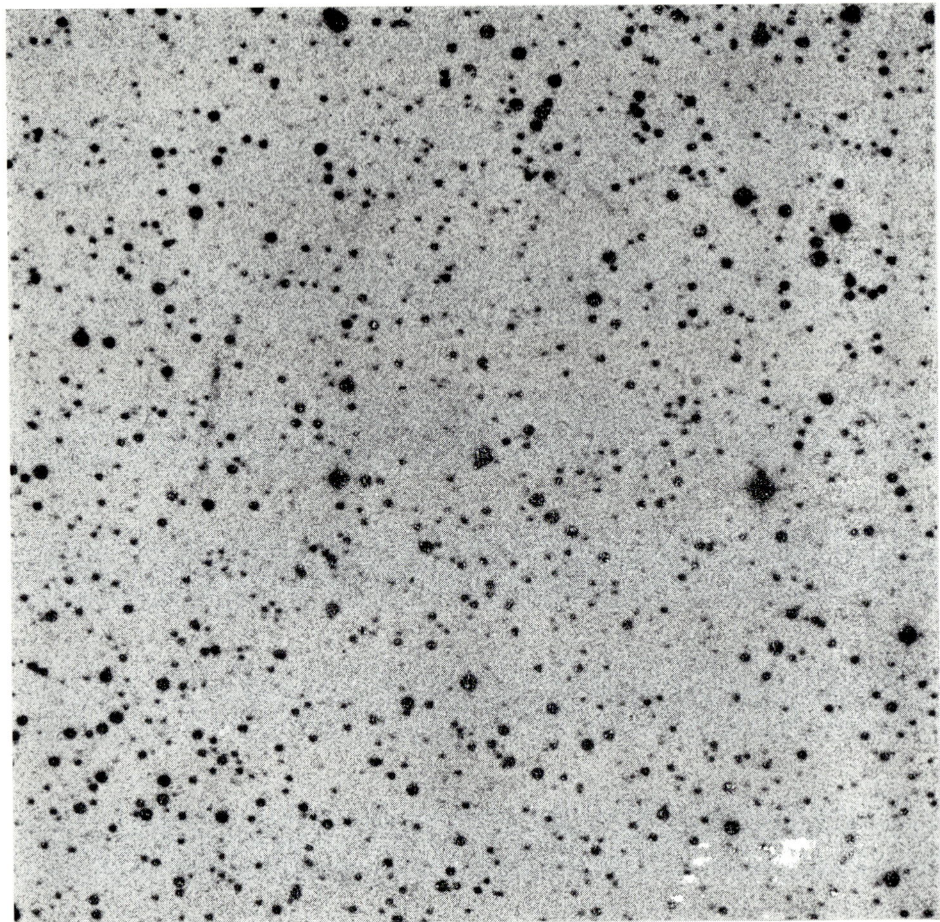

Fig. 2 Photograph of the remnant of Tycho's supernova with the 200-inch telescope in the red (6400–6600Å). Faint filaments can be seen in the left center and upper right of the photograph. The visible details are marked in Fig. 3.

The remnant of Tycho's supernova was found only about ten years ago after radio observations had given a reliable position of the object. It turned out that Tycho's position was off by a few minutes of arc, which is perhaps excusable in view of the fact that this was the first position that Tycho ever determined. Optical observations (Fig. 2) show some faint filaments, and there is some exceedingly faint diffuse nebulosity which needed several plates to confirm. It is obvious that one cannot find out very much about the remnant from the optical information. However, we now have a radio brightness distribution determined at Cambridge with the "one-mile" aperture synthesis telescope. This is shown in Fig. 3. Tycho's supernova can be

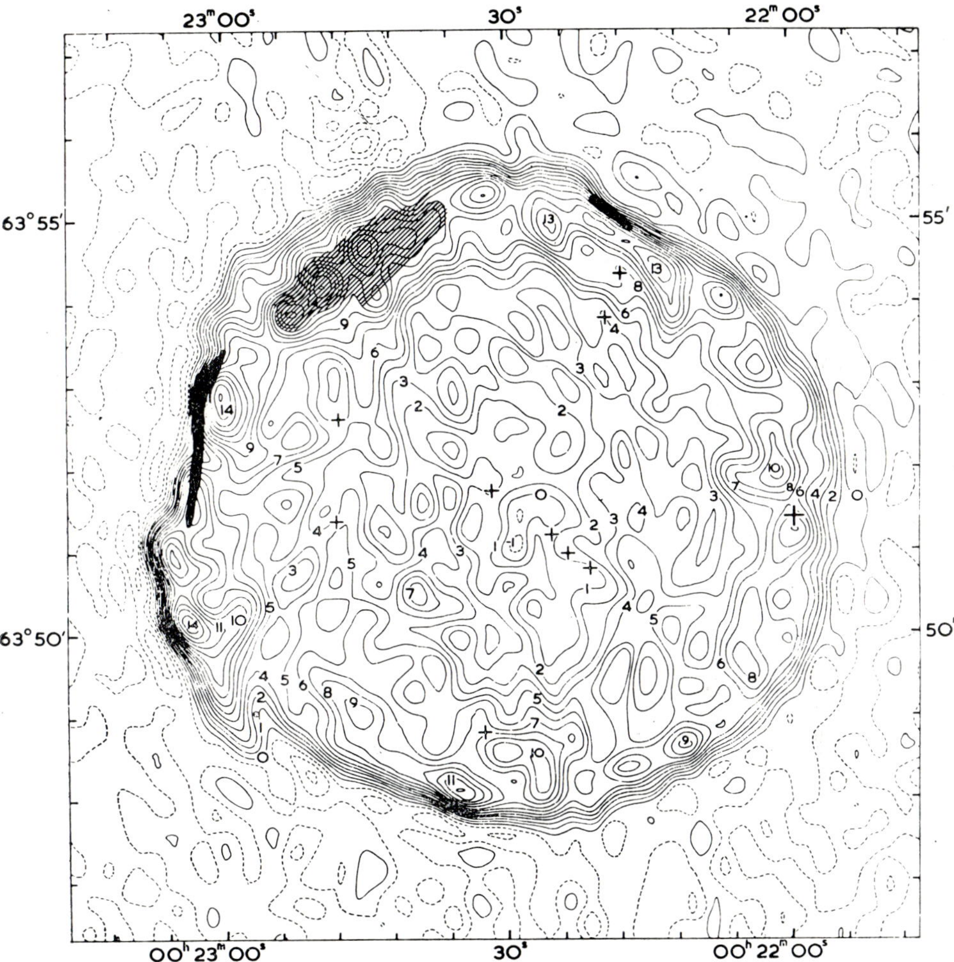

Fig. 3 Remnant of Tycho's supernova, observed at 1407 MHz with the one-mile radio telescope at Cambridge. Optical features are indicated by black and hatched areas. (J.E.Baldwin in IAU Symposium No. 31, *Radio Astronomy and the Galactic System*, p. 352, 1967. Copyright International Astronomical Union.)

seen as a spherical shell bounded by a very narrow ridge. The center of the shell agrees very closely with the location of the center that I previously estimated on the basis of the optical details. The angular diameter of the remnant is about 3.7 minutes of arc. The distance to the remnant has been determined to be 5000 pc by Menon and Williams on the basis of 21-cm observations. Then the radius of the shell becomes 5.4 pc. The distance enables us to find out the brightness of the supernova at maximum. In order to do this we must know the interstellar absorption, which can be determined from the color of the supernova around maximum. Tycho and Kepler were not able to do B-V photometry, but they did make some clear statements about the color being similar to that of some planet or star. We can translate this information into B-V values; we then find that the interstellar reddening is about 0.7 magnitude. This gives an interstellar visual absorption of 2.1 magnitudes, which agrees reasonably well with the results of photometric observations in the region. We then find that the visual magnitude at maximum was -19.6, which agrees very well with the average for Type I.

From the known radius and age we can find the average velocity of expansion, which turns out to be 13,400 km/sec. We must assume, of course, that interaction with the interstellar medium has slowed down the expansion so that the initial velocity must have been considerably higher—perhaps 20,000 km/sec or more. Such values are quite acceptable. Figure 4 shows two spectra of the supernova in NGC 1003. Discovered in 1937, it is one of the few supernovae for which we have information before and at maximum. The enormously wide bands in the center of the spectrum are consistent with an expansion velocity of the order of 20,000 km/sec.

There is one question which is now easily understandable; namely, why is the Tycho remnant so faint? This is in fact typical of all Type I remnants. If a shell is expanding in the interstellar medium with a velocity of the order of 20,000 km/sec. temperatures of the order of $10^9\,°K$ will result. At such enormous temperatures the optical emission is very weak. The reason why we do not see these remnants is that they are too hot to be seen. If we look at the spectrum of the Tycho remnant we find that the radio emission follows a simple power spectrum to about 10,000 MHz, with a spectral index of -0.67. There is negligible optical emission, but x-ray emission is observed. The data on the energy emitted in the various spectral regions are summarized in Table I on page 42. For lack of anything better one can apply the similarity solution for a strong explosion given by Sedov; one then finds that the initial energy of the explosion was about 10^{51} ergs. If we take a third of this value as in-

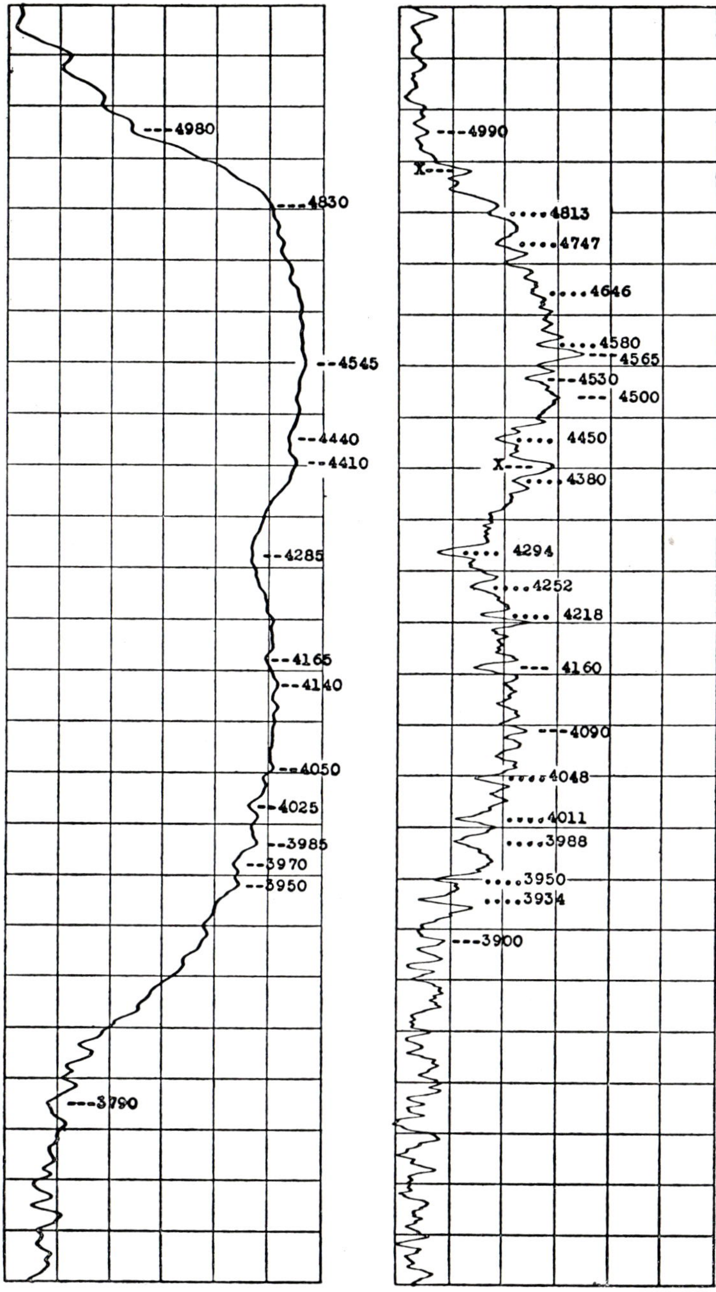

Fig. 4 Two spectra of the supernova in NGC 1003, two days before maximum.

dicative of the kinetic energy and use 20,000 km/sec as the initial velocity we then find that the ejected mass must be of the order of 0.1 solar mass. This value is quite reasonable, but is not necessarily correct.

For Kepler's supernova data are much poorer; the remnant lies in a direction such that 21-cm measurements cannot provide the distance or the ambient density of the interstellar matter. In addition, we have no good observations of the size and structure of the remnant—it lies too far south for the Cambridge observations. If we assume that it had average brightness we calculate a distance of 6700 pc, and we obtain in any case the same picture of a very high average velocity. The overall picture is essentially consistent with that of Tycho's supernova. There is one other object which probably was a supernova of Type I, and that is the supernova of 1006. There is a good radio observation of the remnant; a search by Baade in the red, blue and ultraviolet has revealed nothing optically, which agrees with our explanation of the very low optical brightness. The data on the outburst are very doubtful and open to interpretation. One can say only that it fits into the same general type of scheme.

We come now to the Crab Nebula. It is now completely clear that the Crab Nebula cannot be a remnant of an ordinary supernova of either Type I or II which exploded in an ordinary region of the galaxy. This follows very simply from the fact that the Crab Nebula is a very strong radio source and is also considerably brighter optically than Tycho's or Kepler's supernovae. We can state quite definitely that there is no other object like the Crab Nebula in our half of the galaxy. Any object similar to the Crab Nebula not farther away than the galactic center would be a strong radio source at a level at which all other radio sources have been identified or clearly recognized as thermal emitters. In addition there is the fact that the Crab Nebula emits very strong synchrotron radiation in the visible spectrum, with considerable activity which seems to indicate that some process of injection is still going on. The Crab Nebula is undoubtedly the result of a rare event. One may believe that it was an ordinary supernova that by chance went off in some region where there are most unusual interstellar densities or magnetic fields but that is an ad hoc assumption which I feel one should not make.

This leads us to reconsider whether the event of 1054 was in fact a Type I supernova. The information about that event is rather marginal. We know that 23 days after maximum the apparent magnitude was -3.5. At that point the supernova ceased to be visible in daytime. This is a fairly reliable value. The other datum is that at 650 days the object ceased to be visible. This was originally interpreted to mean that it had fallen below 6th magnitude. However, if we consider where the Crab Nebula was with respect to the sun at the time when it could no longer be seen, we find that it was quite close to the sun. After sunset it was at an altitude of less than 20° above the horizon, and until an observer is convinced that he can no longer see it the object could be even lower. Thus the object was at a very low latitude; it was in a part of the sky close to the sun where the zodiacal light is bright; so one can assume that it must have been considerably brighter than 6th magnitude. There is one complication in that there are some data on Chinese star counts which suggest that the Chinese were able to see stars fainter than 6th magnitude. However, I think it is not unrealistic to say that it was 5th magnitude or brighter when it disappeared, so that the drop in magnitude was 8.5 magnitudes or less during the time that it was observed. How does this compare with the decline of a typical Type I supernova? There has only been one modern supernova which has been observed for an equivalent length of time. This was IC 4182, for which Baade has a photographic light curve extending 635 days. Baade's photographic scale in 1937 was off at the faint magnitudes, and if this is corrected for, one obtains a drop of 10.6 magnitudes. This value is consistent with those obtained by extrapolation of the light curves of Tycho's and Kepler's supernovae. Comparing with the result for the Crab Nebula, we see that the agreement is not so convincing that we can only conclude that it must have been a supernova of Type I, and it is not sufficiently different to say it could not have been. Consequently, we can draw no conclusion, and additional evidence must be had. One could also question the evidence as to whether the Crab Nebula is really the result of the event of 1054. I will not go into that question. One could discuss this problem endlessly, without reaching any definite conclusions. It is sufficient to say, however, that what we know of the Crab Nebula itself fits into the picture that it is not a remnant of a supernova of Type I.

The Crab Nebula is roughly an elliptical object. We do not now whether it is prolate or oblate, and we must know this if we want to determine the distance by comparing the transverse expansion with the radial velocity. Baade tacitly assumed that it was oblate. There was probably a psychological factor involved. At that time the extragalactic distance scale was wrong; the extragalactic supernovae were considered to be much closer with a consequent error of about 5 magnitudes less in brightness. The Crab Nebula on this scale became enormously superluminous, and if Baade had made it prolate the situation would have been even worse. At any rate, Woltjer first called attention to the fact that the discussion of the relative intensities of the forbidden lines of O II and O III runs into difficulty unless the distance is larger than what was assumed, but this difficulty would be removed if the Crab Nebula were prolate. That it is prolate is physically much more reasonable, and this has now been generally recognized. If we consider the Crab Nebula to be prolate, then the distance becomes 1800 pc and the velocity of expansion in the direction of the major axis becomes about 1700 km/sec. Assuming this distance and taking interstellar absorption into account, we obtain an absolute visual magnitude at maximum of -18.2. This is somewhat faint, but not too faint for a Type I supernova. On the basis of the observed expansion, we find that this expansion started at most slightly later than the event of 1054. The expansion age thus indicates that there may have been some acceleration. There is certainly no room for deceleration. Consequently we cannot explain the difference between the observed velocity of expansion of 1700 km/sec and that of 20,000 km/sec expected for a Type I supernova by assuming that the expanding gas was slowed down by interaction with the interstellar medium.

One can also argue that there has been no deceleration on the basis of mass considerations. The mass of the Crab Nebula is not well known. One determination has been made by O'Dell. The procedure is to determine electron density and electron temperature and thus the volume emissivity from emission line data; from the total emission in hydrogen lines one can calculate the volume; multiplying the volume by the electron density gives the mass. When O'Dell's results are corrected for the larger distance, the mass is found to be about 1.6 $M_\odot$. This is the mass in the filaments and thus is a lower limit, but the mass in the synchrotron system is likely to be small. On the other hand, Osterbrock obtained the mass by measuring the electron density and making an estimate of the size and number of filaments. His value, corrected for the larger distance, is 0.34 $M_\odot$. If the electron density is made larger, O'Dell's value will decrease, while Osterbrock's value will increase. If the electron density is low by a factor of about 2.2, the two values agree and we get a mass of about 0.75 $M_\odot$. It seems reasonable, therefore, to think of the filamentary system of the Crab Nebula as having a mass of about 1 $M_\odot$. One can also estimate the interstellar mass which has been swept up. This is somewhat difficult because the Crab Nebula is in a direction where we cannot determine the interstellar distances and densities with 21-cm radiation. However, one can extrapolate from the surrounding regions in the sky, and one obtains an interstellar mass of the order of 0.1 $M_\odot$. This is clearly not adequate to decelerate the velocity from 20,000 km/sec to 1700 km/sec.

Finally, additional supportive evidence is provided by the observation of an abnormal hydrogen-to-helium ratio in the spectrum of the Crab Nebula. When we compare the spectrum of the Crab Nebula with spectra of planetary nebulae in various stages of excitation, we see that the relative intensities of the spectral lines cannot be explained without assuming that there is more helium than the average. Determinations made by Woltjer lead to the result that the helium-to-hydrogen ratio is about three times higher than the average interstellar value. This evidence that there is an abnormally high He/H ratio is a strong indication that the mass of the Crab Nebula cannot be swept-up interstellar mass.

It seems clear that there was no deceleration, and so it seems obvious that the Crab Nebula did not result from a Type I supernova. I might point out that the only kind of supernova where a velocity of a similar order of magnitude as in the Crab Nebula has been observed is the supernova in NGC 1058 (1961)—Zwicky's Type V—but there is no evidence that the Crab Nebula was that type of supernova, and in fact there is some evidence against this interpretation. It seems to me that the question as to what kind of supernova produced the Crab Nebula must be left open.

Now let us consider the Cassiopeia source. This is obviously an object that is highly unique in the galaxy. The radio emission is so strong that if Cassiopeia A were located anywhere in the galaxy it would still be among the six strongest radio sources in the sky. The data for the Crab Nebula and the Cassiopeia source are compared in Table I. Looking at the spectral indices, we note that the radio spectrum of the Crab Nebula is considerably flatter than that of Cassiopeia A. For the Cassiopeia source no continuous optical spectrum is observable; however, the absorption is high. Extrapolation of the radio spectrum indicates that in the absence of obscuration we would expect optical synchrotron radiation at a level of about half the average brightness of the night sky. The x-ray emission is observed to be somewhat less than what a straightforward extrapolation would predict, which means that as in the Crab Nebula the spectrum cuts off somewhere in the x-ray region. For Cassiopeia A the distances determined from 21-cm radiation and from expansion are in agreement; the accepted value is 3400 pc. The optical region is roughly spherical with a radius of 2 pc. There is one remarkable feature; on the eastern side of the object there is a faint flare which extends out to about 3.8 pc. We now have an excellent determination of the brightness distribution from the Cambridge "one-mile" aperture synthesis telescope (Fig. 5). The radio shell is not as smooth as the Tycho shell, and con-

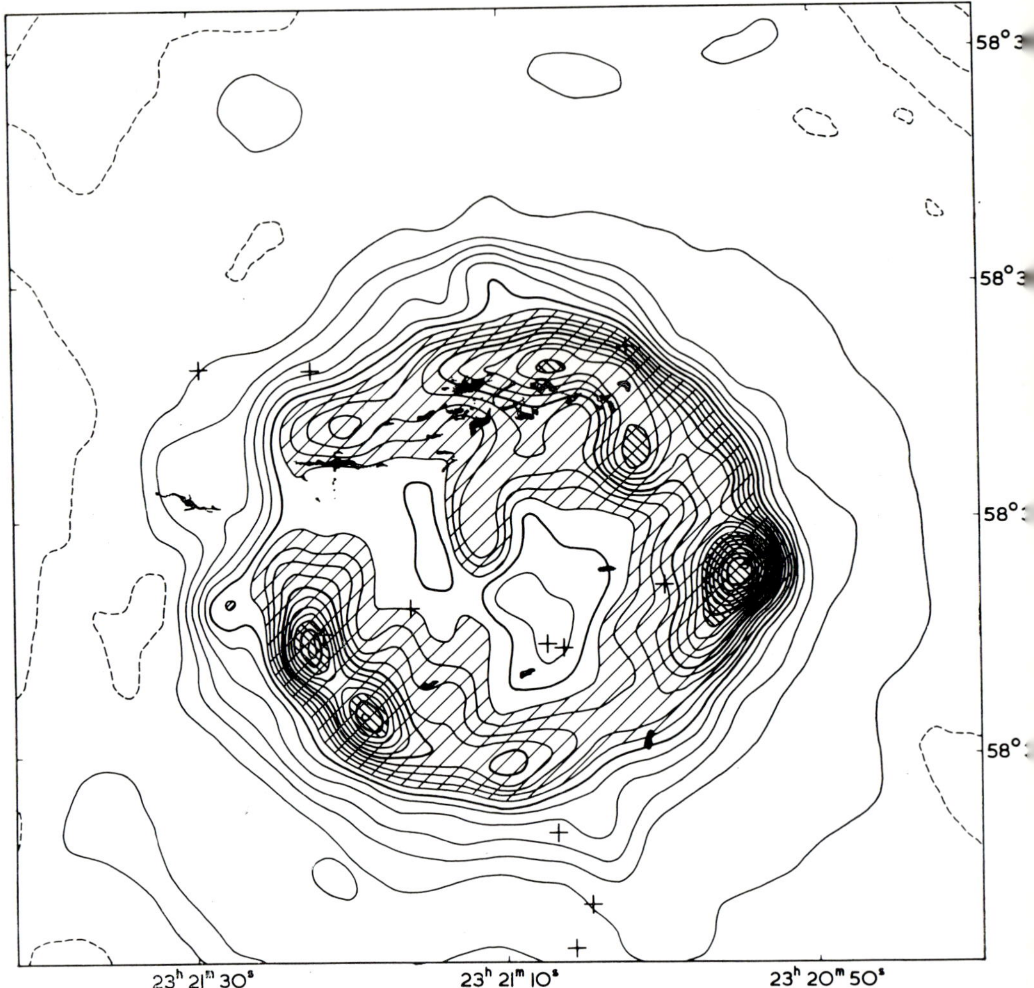

Fig. 5 Radio brightness distribution for Cassiopeia A. The brightest optical details are shown.

tains several condensations. A conspicuous gap in the radio shell corresponds to the location of the flare.

This suggests as one possible explanation that the outburst resulted in the ejection of one shell with high velocity which expanded into the interstellar medium, compressed it, and thus formed the radio source. There is no reason to assume that the interstellar medium is homogeneous over distances of about 5 pc. If in some direction the ejected material encounters low interstellar density, there will be little compression and radio emission; the shell will show a gap. The material ejected in this direction will escape with little deceleration, get farther out and form the flare. We know from the radial velocity observations that at the edge of the optical shell at a radius of 2 pc the velocity is 7400 km/sec. Baade originally assumed that this was almost undecelerated expansion. This is not an unreasonable assumption, since it turns out that the interstellar mass involved is relatively small, while the mass of the filaments is of the order of 1–2 $M_\odot$. Extrapolating backwards one then finds that the expansion began in 1702. However, if we now assume that the shell has been decelerated while the flare went out without deceleration, then the event occurred more recently—probably not more than 100–150 years ago. The major drawback of this picture is that it requires a velocity of the flare of about 30,000 to 40,000 km/sec, which is a velocity never observed in a supernova.

Consequently, we should consider a different picture. It was originally suggested that Cassiopeia A was a remnant of a supernova of Type II, since it did not look like the Crab Nebula which was thought to be a Type I. However, Cassiopeia A does not look like the Tycho remnant either, so it is still possible to consider it to be a Type II remnant. In supernovae of Type II multiple ejection is not infrequent and may in fact be the rule. Let us then suppose that initially there was ejection of a fast shell. As before we assume that this shell was decelerated and formed the radio shell, but that some portion of it escaped in the direction where the source shows a gap and formed the flare. The initial fast ejection was followed by the ejection of a slow shell. The fast shell swept the space clear so that the low shell moved without much deceleration. The date of the outburst then remains 1702. If we assume that the event occurred at that date, we find that we need a velocity of 14,000 km/sec to make the flare The remainder of the fast shell has since been decelerated to form the radio shell. Then the optical shell at 2 pc can be accounted for by an expansion velocity of 7400 km/sec. The overall picture is very similar to what has been found for supernovae of Type III. Poveda has argued, on the basis of the transformation of the spectra of Type II from a continuous spectrum to an emission line spectrum, that the mass ejected by a Type II supernova ought to be quite small. The first crude estimate was that the ejected mass was about 0.02 $M_\odot$. Somewhat better computations lead to a value of 0.1 $M_\odot$. This is at variance with the mass of 1–2 $M_\odot$ that one finds in the filaments. However, Type III supernovae are quite similar to Type II except that they appear to have a much higher mass. Furthermore, a Type III

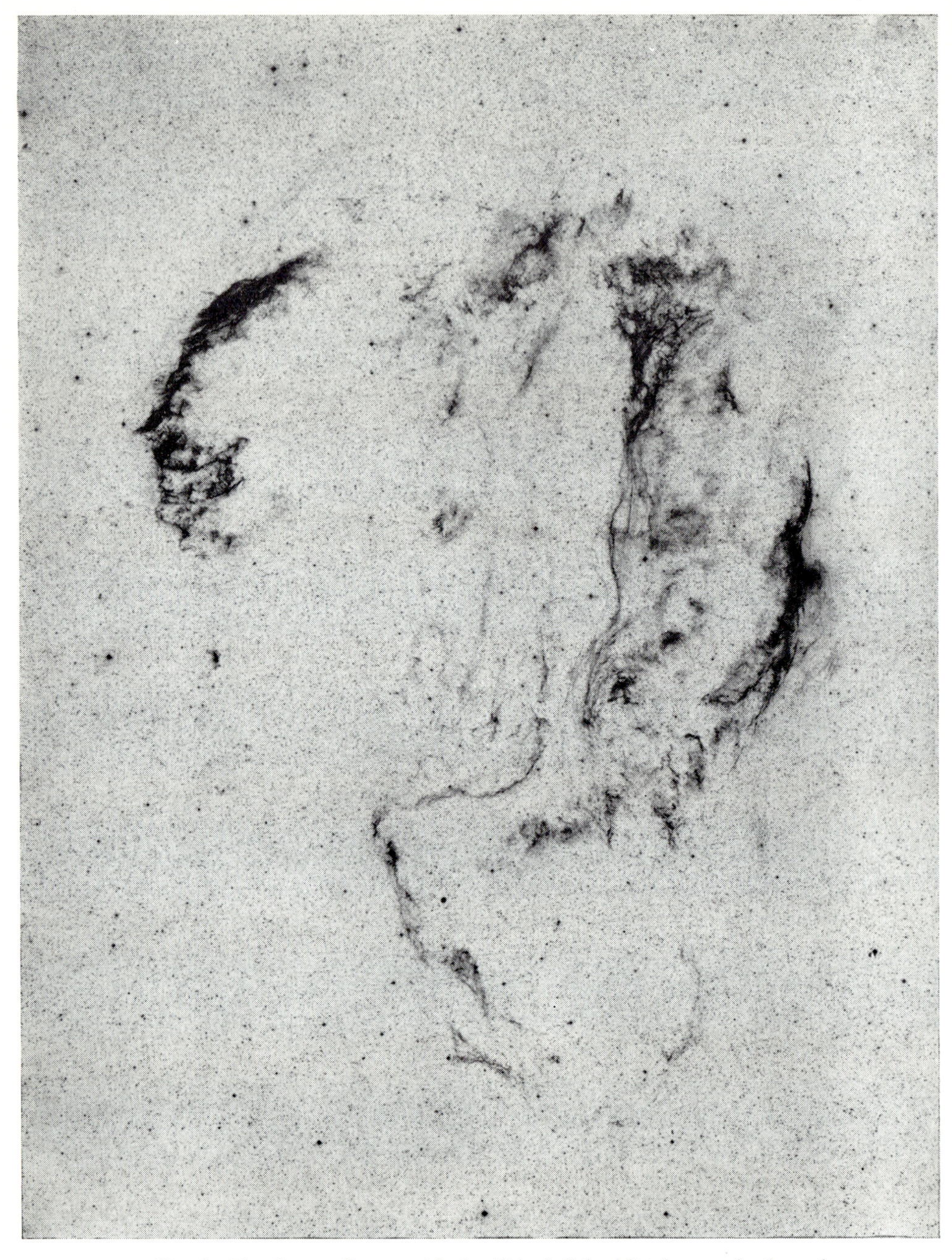

Fig. 6 The Cygnus Loop, with the 48 inch Schmidt telescope in the red (6400–6600 Å). 1 mm = min. of arc.

Fig. 7 Brown & Hazard Nr. 9, with the 48 inch Schmidt telescope in the
red (6400–6600 Å). 1 mm = min. of arc.

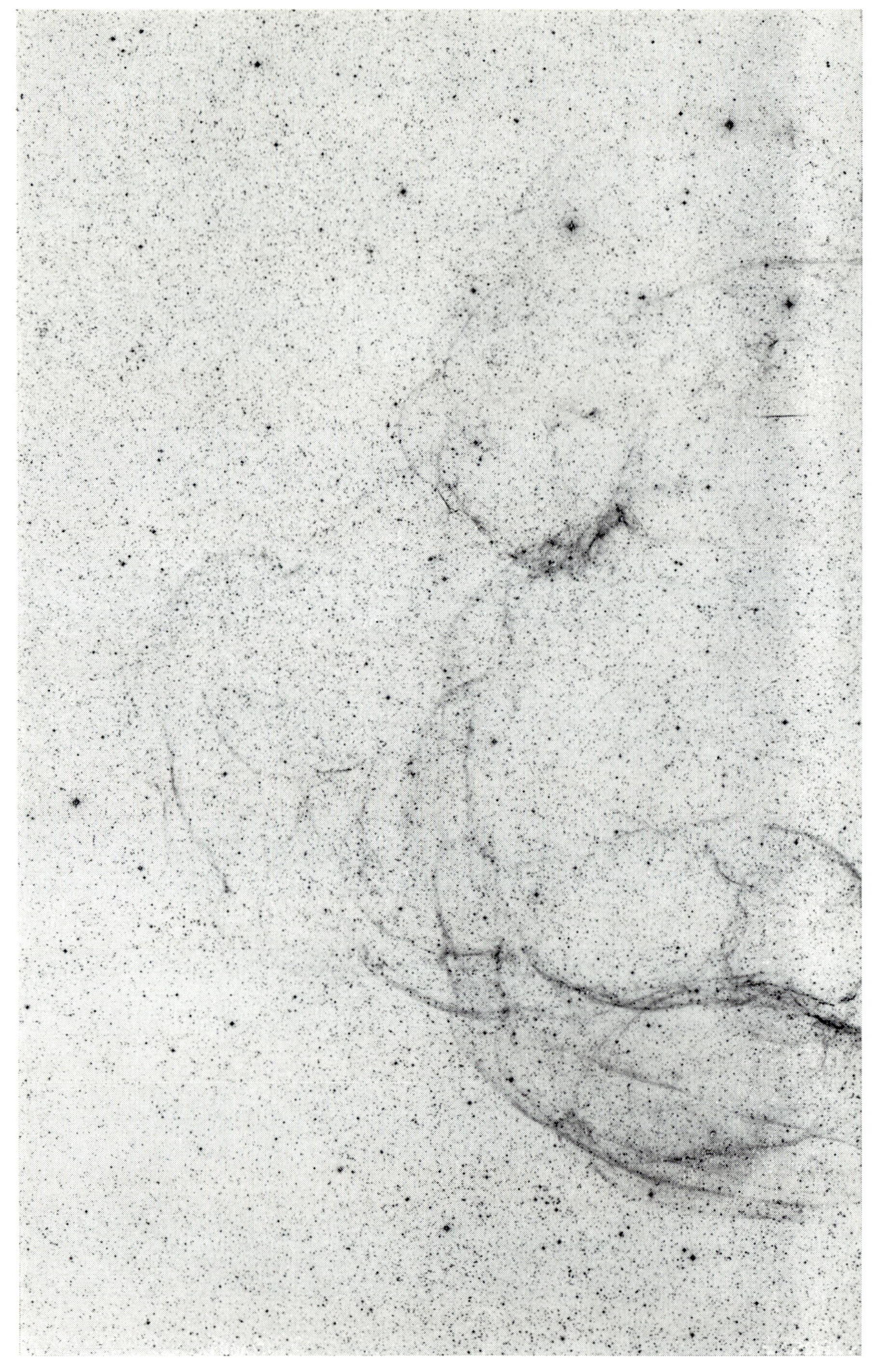

Fig. 8 Shajn 147, with the 48 inch Schmidt telescope

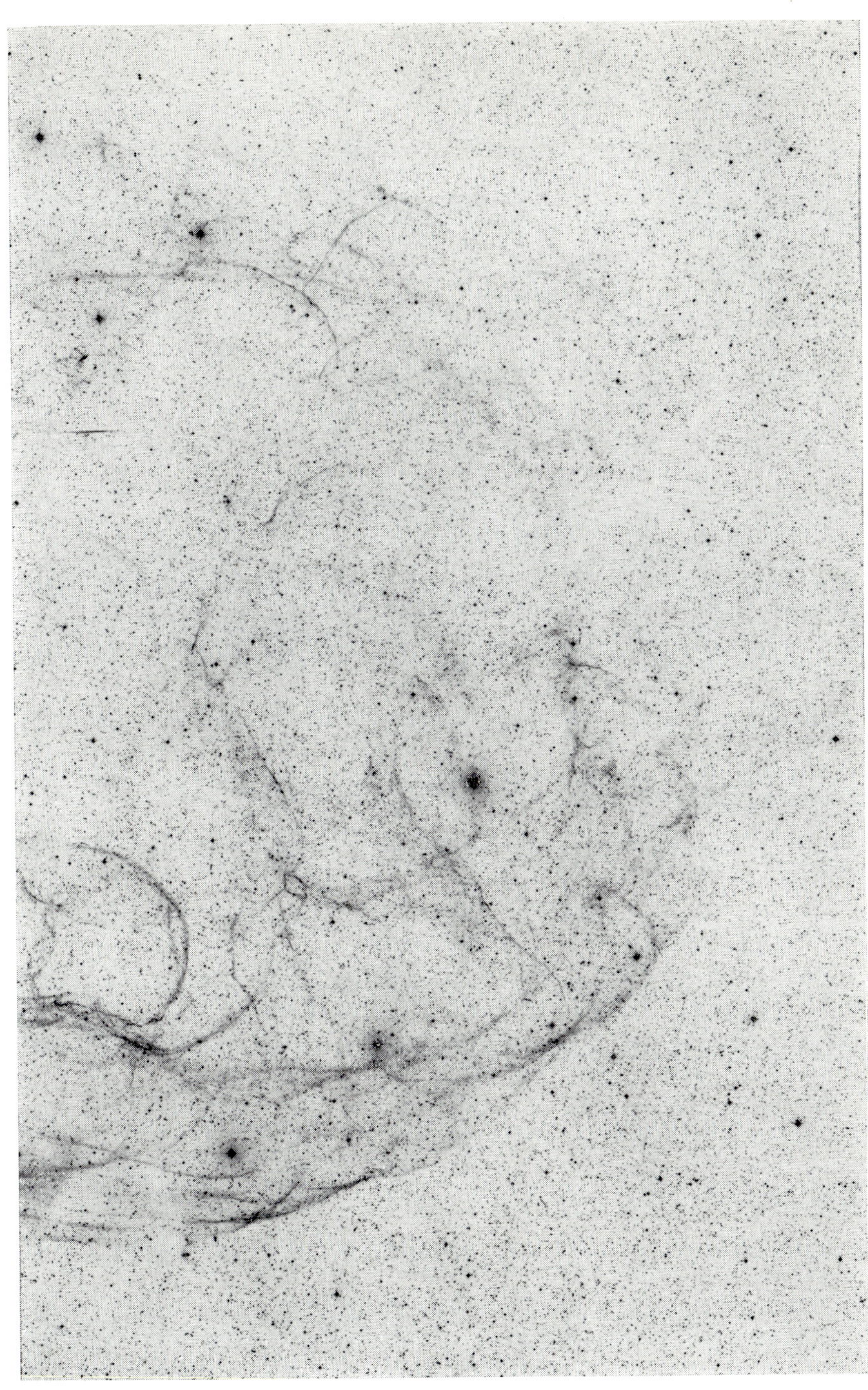

in the red (6400–6600 Å). 1 mm = min. of arc.

supernova is a rare event, so it seems resonable to interpret Cassiopeia A as a Type III remnant.

Thus far we have not identified any remnants as resulting from Type II supernovae. This is rather curious, since most estimates indicate that the frequency of Type II supernovae is not much different from that of Type I. One possible explanation is that much less mass is ejected by a Type II supernova. This would mean that there is much less kinetic energy to begin with, so that Type II remnants might fade out considerably more rapidly than Type I remnants.

There are quite a few large, roughly spherical radio sources which have been found to have relatively low expansion velocities. The Cygnus Loop is the best example of this type of object; it is in fact the brightest one. These objects were ascribed to Type II supernovae years ago on the basis of the argument that the Crab Nebula was a Type I remnant. The low expansion velocities did not seem to indicate enough energy to make these rather large and old remnants. However, we now realize that Type I remnants are as energetic, if not more so, as Type II remnants. This means we can no longer assume as certain that these objects are remnants of Type II supernovae; they could be remnants of either type. These remnants have great individuality. Figure 6 shows the Cygnus Loop, a shell which is still recognizably spherical (a large scale reproduction is in *Sky and Telescope*, **17**, center, 1958). The radio source Brown & Hazard Nr.9 is seen in Fig. 7; it is also roughly spherical but is much fainter with considerably less filamentary structure. On the other hand, Shajn 147 (Fig. 8) is a rather well-bounded object. It is somewhat reminiscent of the radial structure of the Crab Nebula in showing many bulges; there are very long and apparently very sharp filaments. It is very difficult to get optical information for such objects; the Cygnus Loop is the only one for which the distance and size are known.

Whether the variety of appearance among these remnants has anything to do with the variety of supernovae is not clear. The age of the Cygnus Loop is usually estimated to be about 50,000 years. It seems likely that after this time the shape and structure of a remnant depends more on the conditions in the interstellar medium than on conditions of the original explosion.

References

Complete references can be found in: Minowski, R., 1967, *Nebulae and Interstellar Matter*, Eds.
        L.H. Adler and B.M. Middlehurst (Chicago, University of Chicago Press), Chapter 11.

DISCUSSION

P. Morrison   Do these objects tend to lie near the galactic plane?

R. Minkowski   This is very difficult to judge because without the distance we cannot determine how far from the galactic plane they really are. For example, the Kepler remnant does not appear to be very far from the galactic plane. However, if it is at a distance of 6700 pc, then it is 700 pc from the galactic plane, which is reasonably large as z-distances go. On the other hand, the Cygnus Loop is in angular measure rather far from the galactic plane but it is at a distance of only about 750 pc, so that in fact it is not very far from the galactic plane. There is a general tendency for these supernova remnants to be concentrated near the galactic plane, but we cannot tell how close a concentration it is.

S. Colgate   Can you give estimates of the temperature and total bolometric luminosity at maximum?

R. Minkowski   As long as we cannot interpret the spectrum of a Type I supernova, we cannot determine the temperature. We are not even certain that the supernova has at that time a spectrum which permits description by a temperature. There might be so much x-ray emission that the concept of temperature becomes confused. Type II supernovae have in the beginning a continuous spectrum which is that of a blackbody at 40,000–50,000 °K, but this could be quite misleading. Similarly, we do not know what the bolometric correction is. No one has ever observed the spectrum of a supernova in the far ultraviolet or x-ray regions. If there is any intensity in the x-ray region the total energy goes up enormously.

L. Woltjer   Do you have any comments on the very low radial velocities that you observed in the filaments of the Kepler and Tycho remnants?

R. Minkowski   In the Tycho remnant the low velocities are quite consistent with the fact that these filaments are at the outer edge of the object. With regard to the Tycho remnant, it would be interesting to measure the proper motion of these filaments. In the Kepler remnant there is a complication because the clouds have velocities of the order of -200 to -300 km/sec which clearly do not fit into any picture of an expanding shell. You have made the suggestion that these might be interstellar clouds which are excited and possibly accelerated by the outburst and I don't know of any better interpretation. There is some objection in that if the object is 700 pc above the galactic plane one would not expect to find dense clouds there, but on the other hand it is in the inner part of the galaxy. In Cassiopeia A there are low-velocity filaments which are quite sharp and are distinctly different from the faster-moving ones. They have mostly velocities of -100 to -300 km/sec, in the same range as the filaments in the Kepler remnant. Again you have suggested, as I did, that these might be the result of interactions of condensations in the interstellar medium with the expanding shell. I see no real argument against this, but the slow moving condensations are essentially not interpreted. Of course it could be a circumstellar rather than interstellar medium, but if this material was so close to the star I think it should have been accelerated more than what is observed.

R. Macklin   Can the filamentary structure be explained by large-scale turbulence?

L. Woltjer   It's much more likely that it's some sort of Rayleigh-Taylor instability of the remnants in the interstellar medium.

R. Minkowski   We don't even know what the filaments are. Some people like to think that the filaments are really ribbons. It's unlikely that they are large sheets seen edgewise, but they could be ribbons seen edgewise. However, there are some details in the Cygnus Loop for which this model fits very badly. There are bundless of filaments which appear to be twisted around, as on the outside of a cylinder, and if these filaments were ribbons one would expect to see them broaden. Filaments in general are frequent features of astronomical objects and I do not think there is a convincing theory of filaments.

Table I   Data for supernovae remnants

	Tycho	Crab Nebula	Cassiopeia A
Radio Power (10–10^4 MHz)	7×10^{33} ergs/sec	2.1×10^{34} ergs/sec	2.4×10^{35} ergs/sec
Spectral Index	-0.67	-0.26	-0.77
Optical Power	—	2×10^{37} ergs/sec	?
X-Ray Power (1–10 A)	4×10^{37} ergs/sec	4×10^{37} ergs/sec	2×10^{37} ergs/sec
Distance	5000 pc	1800 pc	3400 pc

3

SUPERNOVAE
AND SUPERNOVA REMNANTS

A. Poveda

Observatorio Astronómico
Universidad Nacional de México
Observatorio de Tonantzintla

and

L. Woltjer

Department of Astronomy
Columbia University, New York

Abstract

A catalog of twenty-five supernova remnants is presented. It is probably free of spurious objects. The objects are shown to satisfy a relation between radio surface brightness at 400 Mc/s and radius with quite small scatter. On the basis of this relation, distances can be determined. The accuracy of these distances is sufficient to show that the supernova remnants delineate local spiral structure as clearly as do young galactic clusters. Evidence is presented that remnants in the galactic halo expand faster than those in the disk. Because of this our sample contains mainly Type II supernova remnants. Ten positional coincidences to within $2.5°$ between supernova remnants and X-ray sources are noted in a sample where less than two random coincidences are expected. The X-ray emission appears to be correlated with age, declining more slowly than the radio emission in the early phases.

The relation between surface brightness and radius does not satisfy Shklovsky's theory. A possible explanation of the observed relation is given. The optical spectra of the remnants are discussed and the inadequacy of the usual treatment of collisional excitation is noted. The energies of some remnants are evaluated and are below 5×10^{49} ergs. In the literature the energies released in Type II supernova outbursts are frequently given as of the order of 10^{53} ergs. It is shown that neither the supernova remnants nor the supernova spectra indicate the release of so much energy. We conclude that there is no evidence that supernovae of Type II are catastrophic events related to ultimate stellar collapse.

I Introduction

DISCUSSIONS OF supernova remnants have been given from time to time, but few convincing results have been obtained because of a lack of adequate data. In the past few years, however, many new data have become available, and it thus seems to be worthwhile to investigate these objects anew. Obviously, the first need is to establish a set of criteria by which supernova remnants can be recognized and to obtain a catalog of these objects that is as complete as possible. If the criteria are too stringent, there will be very few objects in the catalog and it will be impossible to obtain meaningful conclusions. If, on the other hand, the criteria are too lax, the number of spurious objects in the catalog will be large, and any real correlations will be masked by the scatter introduced by the spurious objects.

Four historical supernovae, those of 1006, 1054, 1572, and 1604, have been well documented and little doubt exists that their radio remnants have been correctly identified. In addition, we consider 21 other radio sources as rather certain supernova remnants on the basis of the following two criteria. First, the radio radiation is non-thermal. This is evident in most objects because the spectral index, α, is larger than 0.2. Additional objects for which we consider the non-thermal character of the radio radiation established are Vela X, for which significant polarization has been measured and, with perhaps somewhat less certainty, the objects Shajn 147, Shajn 34 and CTB 1, in which the amount of ionized gas evidenced by the Hα radiation appears to be completely inadequate to account for the intensity of the radio radiation. The second criterion that all objects in our catalog meet is that either the distribution of radio emission shows clear, shell-like structure, or that a typical filamentary nebulosity coincides with the radio object.

The relevant data on the objects that meet these criteria have been assembled in Table I. First are given the 1950 equatorial coordinates and the new galactic coordinates for the radio sources, except in the case of CTB 1, where the optical position is given. Next follow the angular diameters (d in minutes of arc) which have been obtained in most cases from the published radio contours. These diameters are not necessarily half-power diameters. If the radio object is a shell, then the diameter has been taken such that it represents the outer diameter of the shell. For Sh 147 and the Cygnus Loop the optical diameters have been taken. In the next two columns follow the spectral index and the flux density at 400 Mc/s in units of 10^{-26} w/m^2/sec/(c/s).

Table I Data for supernova remnants

Consecutive columns give the objects, 1950 equatorial coordinates, new galactic coordinates, angular diameter in arc minutes, spectral index, flux density at 400 Mc/s in units of 10^{-26} w m^{-2} sec^{-1} (c/s)$^{-1}$, surface brightness in units of $(4/\pi)$ flux units per square arc minute, distance in pc, height above the galactic plane in pc and radius in pc. Distances marked with an asterisk have been determined independently; the other distances have been obtained from the Σ-d diagram.

Source	α	δ	l^{II}	b^{II}	d	α	S_{400}	Log Σ	D (pc)	x (pc)	R (pc)
CTA 1	00^{h}03^m	72°04′	119°4	9°8	120′	0.2	50	−2.46	2000	+340	36
TYCHO S.N. 1572	00 22.8	63 51	120.1	1.4	7	0.67†	100	+0.30	3500*	+85	3.6
HB 3	02 15	62 30	132.8	1.5	110	0.7	70	−2.23	1700*	+45	28
HB 9	04 57.5	46 36	160.5	2.8	140	0.3	200	−2.00	1100	+55	23
OA 184	05 13.4	41 45	166.0	2.2	70	0.5	16	−2.50	3400	+130	36
VRO 42.05.01	05 23.1	42 49	166.2	4.3	50	0.3	8	−2.50	4700	+350	35
CRAB NEBULA (S.N. 1054)	05 31.5	21 59	184.3	−6.2	5	0.26†	1200	+1.68	1300*	−140	1.0
Sh 147	05 36	27 06	180.8	−2.2	200	0.0	110	−2.56	1200	−45	36
Sh 34	06 05	16	194	−2	180	—	[100]	−2.52	1300	−45	35
IC 443	06 14.6	22 43	189.0	3.2	45	0.41	250	−0.90	1500*	+85	10
PUP. A	08 21.3	−42 52	260.4	−3.3	40	0.61†	220	−0.86	1500	−85	9
VELA X	08 32.5	−45 35	263.9	−3.3	140	0.0	[1000]	−1.3	600	−35	12
VELA Y	08 43.5	−43 42	263.6	−0.6	120	0.4	200	−1.86	1200	−10	22
P 0902-38	09 02.4	−38 29	261.9	5.5	44	0.3	12	−2.20	4200	+400	28
P 1439-62	14 39.1	−62 15	315.4	−2.3	52	0.43	50	−1.74	2300	−95	18
P 1459-41 (S.N. 1006)	14 59.6	−41 42	327.6	14.6	30	0.6	20	−1.66	4000	+1000	18
P 1548-55	15 49.0	−55 59	326.3	−1.7	45	0.26	180	−1.05	1600	−50	11
P 1613-50	16 13.6	−50 55	332.4	−0.4	7	0.6	45	−0.05	4300	−30	4.5
KEPLER S.N. 1604	17 27.7	−21 25	4.5	6.9	3	0.64†	34	+0.57	10,000*	+1200	4.4
W 28	17 58.5	−23 20	6.6	−0.1	30	0.35†	450	−0.31	1300	00	5.9
3C 392	18 53.6	+01 14	34.6	−0.4	22	0.44†	290	−0.22	1600	−10	5.3
HB 21	20 44	50 30	89.0	+4.7	130	0.34†	200	−1.94	1100	+90	21
CYGNUS LOOP	20 48	30 12	73.6	−8.6	170	0.47	300	−1.98	800*	−120	20
CAS A	23 21.2	58 32	111.7	−2.1	4	0.77†	6400	+2.61	3400*	−125	2.0
CTB 1	23 57.8	62 05	117.1	+0.1	35	—	13	−1.96	4400	+10	23

Values marked with a dagger have been taken from Kellerman's (1964) compilation; the others have been determined by other authors or by us. In all cases we have made certain that the quoted values are given in the system

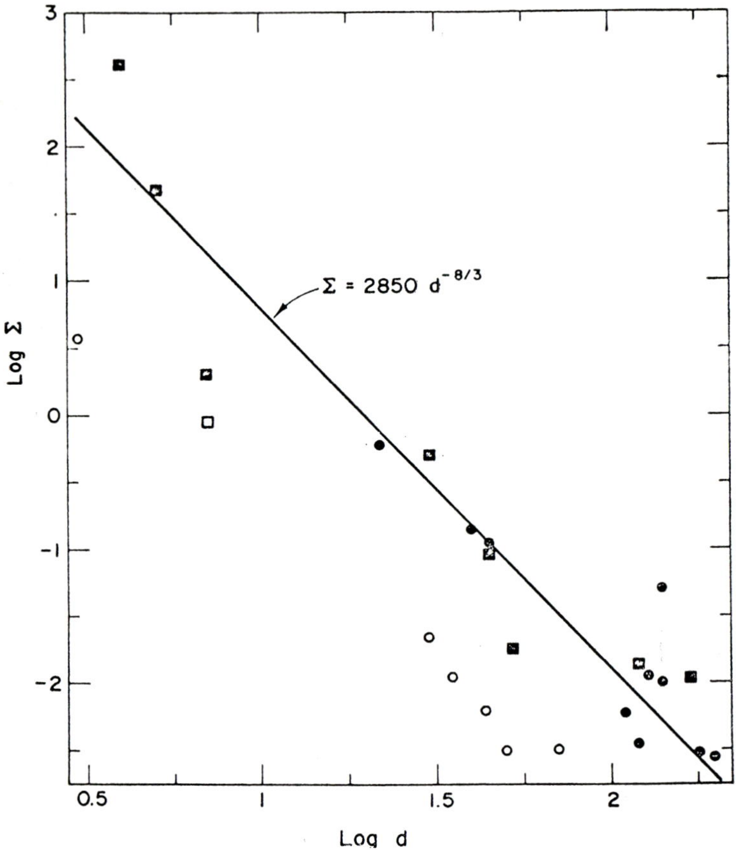

Fig. 1   The $\Sigma - d$ diagram [surface brightness at 400 mc/s in $(4/\pi)$ flux units per square minute of arc]—[angular diameter in minutes of arc]. According to the discussion in the text, objects situated on the line Σ $2850\, d^{-8/3}$ have a mean distance of 1390 pc.

established by Conway, Kellerman and Long (1963). The spectral indices quoted in two decimals should generally have accuracies of ± 0.05, except for the Cygnus Loop, where ± 0.10 may be a better estimate. Those quoted in one decimal place have uncertainties of ± 0.1 or ± 0.2. The next column gives $\log \Sigma$, the logarithm of the surface brightness in $(4/\pi)$ flux units/square min of arc. The remaining columns give the distances, the altitudes above the galactic plane, and the radii of the objects. They will be discussed later.

II   The $\Sigma\text{-}d$ Diagram, Distances, Galactic Distribution

After trying some alternative possibilities, we found it most useful to represent the radio data for the supernova remnants in a diagram in which $\log \Sigma$ is plotted against $\log d$ (Fig. 1). The main advantage of this representation is that Σ is independent of distance, while d depends on the distance only to the first power. When the diagram is inspected, it appears that a strong correlation between Σ and d exists for the points representing sources with flux densities larger than fifty units. The scatter is surprisingly small.

This scatter is due to two factors. First of all, the intrinsic relation between Σ and the true radius R can hardly be perfect, and any scatter around the mean relation will appear in the diagram. Secondly, even if a perfect relation between Σ and R existed, scatter would be introduced into the diagram by the differences in distance between the various sources, which would shift the points to the left or to the right. We will show that a large part of the scatter is due to the second cause, and that the relation between surface brightness and radius is quite precise. Let us consider that we have a set of objects that would satisfy a perfect $\Sigma\text{-}R$ relation. If the objects are confined to the galactic plane, the number of objects closer than D is proportional to D^2. Suppose we have a set of objects that satisfies a sharp selection criterion so that no objects beyond a distance D_0 can be observed. It can easily be shown that the mean distance of all objects observed will be $0.75D_0$, and that the root mean square scatter around the mean will be $0.20D_0$. If the selection effects are not so sharp, and some sources are observed beyond D_0, the average distance is somewhat larger, and the scatter increases proportionally even more. Consequently, the scatter in $\log D$ should be at least ± 0.15, and the same scatter would be expected in $\log d$.

If, in Fig. 1, we restrict ourselves to the points corresponding to the sources with $S_{400} > 50$ flux units and with $\log \Sigma < 0$, it appears that the scatter in $\log d$ is hardly larger than 0.15. Consequently we are inclined to believe that most of the scatter diagram is, in fact, due to distance effects. The intrinsic relation between Σ and R is apparently very precisely satisfied by most, if not by all objects for which $\log \Sigma < 0$. Fitting a straight line to the points, with $S_{400} > 50$ f.u., we find that $\Sigma \propto R^{-8/3}$.

Having obtained the mean relation between Σ and R, we can now obtain individual distances by measuring the horizontal shifts between the line and the point representing a particular object. In this way, obviously, only relative distances can be obtained and our first task will be to calibrate the diagram by finding the distance to which the mean curve corresponds. Three of the remnants in the lower part of the diagram have independently determined distances. For the Cygnus Loop, Minkowski (1958) measured a distance of 770 parsecs by comparing the radial velocities and proper motions of filaments in the shell. For HB 3 Caswell (1967) found indications of interaction with the neighboring object W 3, for which a distance of 1700 parsecs was estimated from data by Sharpless (1955). IC 443 appears to show clear evidence of interaction with the neighboring nebula Sharpless 249 (Sharpless 1959). This nebula probably is part of the I Geminorum association, for which the distance of 1500 parsecs was given by Sharpless (1965). This distance is compatible with the very fragmentary data that are available about radial velocities and proper motions for IC 443 (Minkowski 1958). For the distance of points on the mean $\log \Sigma$–$\log d$ relation, we then find 1210 parsecs from the Cygnus Loop, 1580 parsecs from IC 443 and 1390 parsecs from HB 3. Consequently, we adopt a mean of 1390 parsecs. When the distance corresponding to the points on the mean curve has been obtained, all relative distances can be converted into absolute ones.

We consider the curve to be well determined only below $\log \Sigma = 0$. Above this, very few points are available to determine the relation, and distances have been estimated from independent evidence. For Cas A we have taken Minkowski's distance determination on the basis of the motion of the filamentary shell. For the Crab Nebula, distances of 1100 or 1500 parsecs have been obtained, depending on whether the assumption is made that the filamentary shell is an oblate or a prolate spheroid; we adopt a mean value of 1300 parsecs. The distance to Tycho's supernova was determined by Menon and Williams from 21-cm absorption measurements. For Kepler's supernova, no reliable distance measurement is available, and we have taken the value of 10,000 parsecs, determined from the brightness of the supernova at maximum (Woltjer, 1964). We have made use of the distances obtained to compute the values of the altitude above the galactic plane and of the radius of the remnants. All these data are given in the last columns of Table I.

On extrapolating the Σ–R relation linearly, we noticed that some of the brighter supernova remnants fall rather close to it and thus the relation between Σ and R possibly remains valid up to the Σ corresponding to the Crab Nebula. Distances for the objects for which independent distance determinations are available were computed on the basis of this Σ–R relation. A comparison of all observed and computed distances is given in Table II, which indicates a very good agreement. For Kepler's supernova all estimates are very crude. Cas A, by far the youngest of all, does not fit to an extrapolated curve, which indicates that for very young sources the situation may be more complex. We will return to this in the next section.

Table II Independently determined distances D_0 compared with distances D_c determined from the Σ–d diagram. The objects that were used to calibrate the Σ–R relation are marked with asterisks.

Source	D_0 (pc)	D_c (pc)
Tycho *SN*	3500	2900
HB 3*	1700	1700
Crab Nebula	1300	1500
I C 443*	1500	1300
Kepler SN	10,000	5300
Cygnus Loop*	800	900

On inspecting the distributions of z-distances, we note only five sources with z-distances in excess of 150 parsecs. One of these is Kepler's supernova, which occurs in our sample only because the supernova event was well recorded. The mean $|z|$ of the twenty-one objects below 300 parsecs from the plane is 64 parsecs. Thus most of the remnants in Table I have a z-distribution like 0- and B-stars. If we confine ourselves to the objects brighter than 50 flux units, we find one object with a z-distance in excess of 150 parsecs and seventeen below this level. In Sb and Sc galaxies roughly comparable numbers of Type I and Type II supernovae are observed. If Type I supernovae are taken to represent a rather old population (perhaps like planetary nebulae?) it would be expected that the average z is of the order of 300 parsecs. Thus the remnants observed in the galactic plane should be mostly Type II; and we conclude that our sample is seriously deficient in remnants at larger z-distances. This may be partly a matter of observational selection, but it is unlikely that the whole effect can be explained in this way. The more probable explanation is that the remnants higher above the galactic plane have a

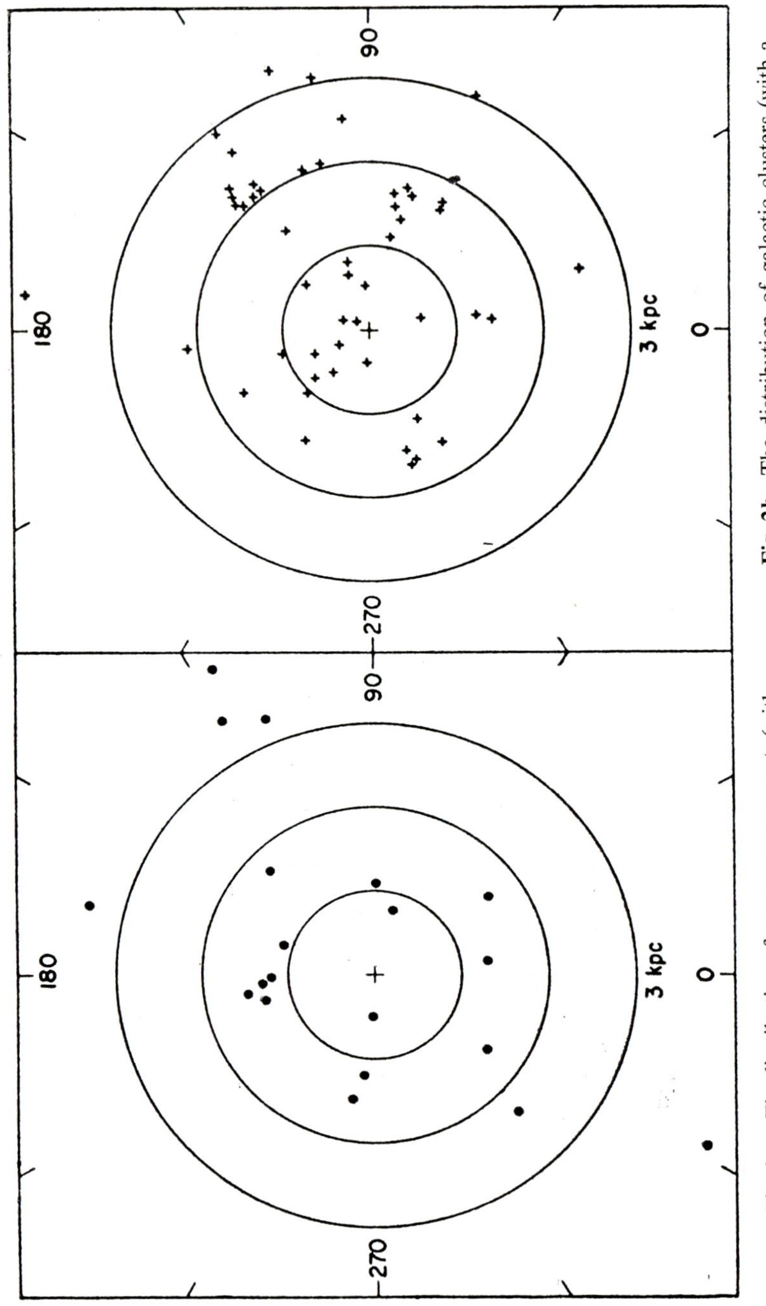

Fig. 2a The distribution of supernova remnants (with $|z| < 150\,\mathrm{pc}$) in the galactic plane. The sun is at the center and galactic longitudes are indicated. The outer circle has a radius of 3 kpc.

Fig. 2b The distribution of galactic clusters (with a main sequence extending to B 3 or earlier) in the galactic plane as taken from Kraft (1965).

shorter lifetime because they expand faster in the tenuous halo medium. The supernova of 1006 would appear to confirm this because it has already achieved a large radius despite its low age. Of course, the different environment of the high z-objects also makes it uncertain that they follow the standard Σ-R relation, which has been calibrated only for objects in the galactic gas layer. We return to this question later.

Next we inspect the distribution in the galactic plane of the twenty objects with z smaller than 150 parsecs (Fig. 2a). It appears to us that there is no question that galactic features (spiral arms or arm links) are clearly delineated by the remnants. In Fig. 2b we plot for comparison the distribution of the clusters containing stars of spectral type B 3 or earlier, for which accurate photometric distances are available. This material has been taken from Kraft (1965). The qualitative resemblance of the two distributions, both of which show the presence of three galactic features, is striking. In our view this constitutes a very direct confirmation of the essential correctness of the distances that we have derived. Perhaps the quantitative correspondence of the two distributions could be improved by changing the calibration of our distances somewhat, but the available data are insufficient to justify such a procedure.

III   Supernova Remnants as X-ray Sources

The Crab Nebula was the first X-ray source to be positively identified as an optical object. Subsequently, Friedman and his associates identified X-ray sources with Cas A and with Tycho's supernova. A source was also found near the position of Kepler's supernova, but this source has not been confirmed by latter observations (Clark *et al.*, 1965).

In Table III we give a list of all positional coincidences between X-ray sources in the catalog of Friedman *et al.* (1967) and supernova remnants, in which the difference of the stated positions is less than 2°5. Three additional X-ray sources for which positions were measured by Chodil *et al.* (1967) were also inspected and this added Vela Y to the list. According to Friedman *et al.* (1967), the object 3C 392 might be identified with Aquila XRI, but because the positional difference is 5°, no strong case can be made. We note in passing that the strong southern radio source 13 S 6 A, which is regarded by Shklovsky as the remnant of the Supernova of 185, is within 3°1 of the position of the rather intense X-ray source Centaurus XR II (described by some as an object in Crux) given by Chodil *et al.* Not enough is known about the

Table III Probable identifications of X-ray sources and supernova remnants. Statistical arguments indicate that two or three of the identifications may be spurious. Consecutive columns give the objects, X-ray identifications, $\Delta\alpha$ and $\Delta\delta$ (the positional differences X-ray source minus supernova remnant), the angular separation of the two, the X-ray flux in counts cm^{-2} sec^{-1} in the interval 1–10 Å and the intrinsic X-ray luminosity under the assumption that one count $= 1.2 \times 10^{-8}$ ergs. Most X-ray data have been taken from Friedman *et al.* (1966). Identifications suggested by Friedman *et al.* are marked with an asterisk.

SN remnant	X-ray source		$\Delta\alpha$	$\Delta\delta$	$[(\Delta\alpha)^2 \cos^2 \delta + (\Delta\delta)^2]^{1/2}$	Counts	L_X (10^{36} ergs/sec)
CTA 1*	CEPH	XR 3	-9^m	$+0\overset{.}{.}8$	$1\overset{.}{.}1$	0.15	0.9
TYCHO SN*	CEPH	XR 1	-8	$+2.2$	2.4	0.3	5
CRAB NEBULA	TAU	XR 1	0	$+0.1$	0.1	2.7	6.5
VELA Y (or X?)	VELA	XR 1	$+8$	-0.3	1.4	[0.5]	1.0
P 1439-62	CENT	XR 1	-11	-0.9	1.5	0.17	1.3
P 1548-55	NORMA	XR 2	-11	-1.0	1.8	0.4	1.5
P 1613-50	NORMA	XR 1	$+10$	0	1.7	0.8	21
W 28	SAG	XR 3	-3	$+1.6$	1.8	2.8	7
CYGNUS LOOP	VUL	XR 1	-10	-1.2	2.5	0.15	0.1
CAS A*	CAS	XR 1	0	0	0.0	0.3	5

radio object, however, to include it in our catalog of remnants. If the identification is correct, it would have interesting implications because the source appears to be variable and its spectrum thermal. Further radio studies of this object are highly desirable.

It is difficult to estimate the number of random coincidences in Table III because of the strongly non-uniform distribution of both X-ray sources and supernova remnants. For the X-ray sources this is a consequence partly of the lack of uniformity in the available surveys, but the main effect for both kinds of objects is due to their strong concentration towards the galactic plane. In the longitude range $l = 280° - 140°$ the NRL survey is far more complete than elsewhere. In this interval there are about 24 X-ray sources with $|b| < 10°$, and twelve supernova remnants. We exclude Kepler's supernova, both from the catalog of X-ray sources and from that of supernova remnants because of its uncertain status. In this interval there are eight identifications within 2°5. The expected number of random identifications in this sample is 1.3. Because of the much smaller number of X-ray sources in the remaining longitude interval, additional random coincides should be negligible.

Of course, with the present positional uncertainties we cannot exclude from this evidence alone the possibility that X-ray sources and supernova remnants are objects with strong spatial correlation, without being identical.

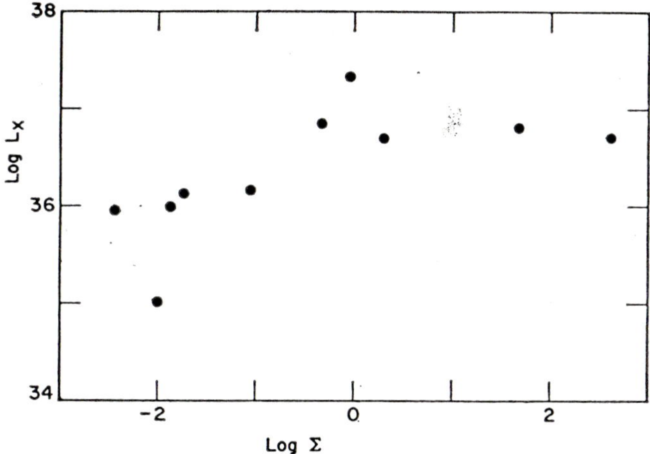

Fig. 3   The intrinsic X-ray luminosity (ergs/sec) versus the radio surface
brightness Σ for the objects in Table III.

For example, if both are concentrated on O-associations or galactic clusters, the observed coincidences could probably also be explained. In either case, however, the distances of the X-ray sources and the associated supernova remnants would be the same. In the last column of Table III we give the X-ray luminosities calculated with the distances given in Table I, assuming that one count is equivalent to 1.2×10^{-8} ergs (Friedman *et al.*, 1967). In Fig. 3 we have plotted these X-ray luminosities against the surface brightness of the remnants, and some correlation appears present. In fact all remnants with Log $\Sigma > 0$ appear to be X-ray sources. Since large surface brightness coincides with small age, we conclude that the X-ray emission decreases slowly with age. The fact that a correlation with age appears to exist indicates that the X-ray sources and the supernova remnants are the same objects.

IV   Evolution and Energetics of Supernova Remnants

Most discussions of supernova remnants have been based on Shklovsky's (1960) analysis. According to Shklovsky, conservation of magnetic flux gives for the time variation of the magnetic field of a supernova remnant

$$H = H_0 \, \frac{R_0^2}{R^2} \tag{1}$$

If during the expansion of the supernova remnant, the relativistic synchrotron electrons maintain an isotropic momentum distribution, the energies of individual electrons vary as R^{-1}. If the electrons have a power law spectrum we find

$$N(E) = KE^{-\beta} = K_0 \left(\frac{R}{R_0}\right)^{1-\beta} E^{-\beta} \tag{2}$$

with $N(E)$ representing the differential electron energy spectrum for the whole remnant. The total synchrotron radiation then satisfies

$$L_\nu \propto KH^{1/2(\beta+1)} \, \nu^{-1/2(\beta-1)} \propto K_0 \left(\frac{R}{R_0}\right)^{-2\beta} \nu^{-1/2(\beta-1)}$$

$$\propto K_0 \left(\frac{R}{R_0}\right)^{-(4\alpha+2)} \nu^{-\alpha} \tag{3}$$

with α the spectral index. With an average spectral index of 0.5 for the remnants in Table I we thus would expect $L \propto R^{-4}$ or $\Sigma \propto R^{-6}$. Instead we find in Fig. 1, at least for $R > 5$ pc

$$L \propto R^{-2/3}, \quad \Sigma \propto R^{-8/3} \tag{4}$$

which deviates strikingly from the predicted expression. On the basis of expression (3) Shklovsky made the prediction that Cas A would decrease its luminosity by 2.0 per cent per year. A remarkable confirmation of the prediction that the source would become dimmer has been made by several radio astronomers (Mayer *et al.*, 1964), but it was found that the actual decline is 1.1 percent per year. The difference may seem slight, but if an R^n law is fitted to the data we find $L \propto R^{2\cdot7}$ instead of the predicted relation $L \propto R^{-5\cdot08}$ (for $\alpha = 0.77$). We note that this variation is much steeper than that of the supernova remnants lower in the $\Sigma - R$ diagram and at its present rate, in a few centuries, Cas A could reach the standard curve.

Shklovsky's discussion applies to a system in which magnetic flux is conserved and in which the total number of relativistic electrons is constant. The conservation of flux becomes inapplicable when interstellar field lines are swept up by the remnant. The number of relativistic particles is not conserved when particle acceleration (or escape) takes place long after the supernova outburst. In the case of the Crab Nebula, it appears that effective acceleration has continued for at least several hundred years after the supernova event (Woltjer, 1958).

The importance of interstellar magnetic fields in the later phases of supernova remnants becomes evident when we compare the radio and H_α brightness distributions in objects like the Cygnus Loop. As noted by previous investigators, a patch of radio emission is closely associated with filamentary structures of NGC 6990; the spectrum shows that the radio emission is nonthermal. At the same time the major maximum of the radio emission in the Loop does not coincide with much nebulosity. Thus, some correlation is present, but the radio emission has a wider distribution than the nebulosities. It seems to us that this kind of behavior is to be expected if most of the field in the remnant is compressed interstellar magnetic field.

As the remnant expands, a shock will propagate into the interstellar medium. For a velocity of 100 kilometers per second the shock will be strong and the medium behind the shock will be mostly ionized; the temperature will be about 200,000 or 300,000 °K and the density will be about four times the interstellar density. A detailed analysis of the hydrodynamical equations would be needed to study the precise properties of the gas flow behind the shock, but in most situations it is a fairly good approximation to treat the gas as isobaric for quite some distance behind the shock front. The gas will radiate (mainly by radiation from collisionally excited states in carbon and oxygen ions) and cool rapidly, while increasing in density. When a temperature of about 10,000 °K is reached the hydrogen recombines, after which the cooling rate drops fast.

We note that while the temperature is high, virtually all hydrogen is ionized. Provisional calculations indicate that the hydrogen recombines only when the temperature is so low that no further collisional excitation of the third and higher levels will occur. Thus, although the ionization is purely collisional, we expect the hydrogen spectrum to closely resemble a pure recombination spectrum. We believe that this is the explanation of the low Balmer decrement that Parker (1967) has reported for the Cygnus Loop, rather than any unseen source of ultraviolet ionizing radiation, as suggested by Parker. A more detailed discussion of the spectrum expected for a shock heated plasma will be reported by one of us later.

In regions where the supernova remnant expands into less dense interstellar matter, an appreciable compression of the interstellar medium still is possible. This leads us to expect only a rather loose correlation between radio and optical features. The correlation is further weakened because the optical radiation may disappear before much decompression occurs after the shock has passed through a dense interstellar region and because not all compressed regions need be accessible to the relativistic electrons of the remnant. Where there is a correlation the radio maxima generally should occur somewhat behind luminous compressed regions which, on projection, appear like thin filaments (Poveda, 1964). An inspection of supernova remnants appears to be in general accord with this expectation.

The relativistic electrons contained in the supernova remnant will gradually diffuse into the magnetic fields swept up in the shell behind the shock. It would be difficult to argue convincingly that they would not diffuse through the shock, but there is no evidence this happens. Perhaps a major diffusion is prevented by the (possibly turbulent) character of the field immediately behiud the shock. As a result of the diffusion process, the density of relativistic electrons in the compressed region may become as large as, but certainly no larger than, the density inside. The emission per unit volume depends on the field strength as $H^{1/2(\beta+1)}$. If the field is squeezed into a layer around the expanding surface, the radiating volume is inversely proportional to the magnetic field strength. Thus, for a given interstellar magnetic field, the total radiation emitted would be proportional only to $H^{1/2(\beta-1)}$ or to H^{α}, with α the spectral index. With a typical α of 0.4, the total emission thus depends only weakly on the compression. If K is the constant in the energy spectrum for the whole remnant, we thus have for the radio luminosity

$$L_v \propto KH^{\alpha}H_0 v^{-\alpha}, \tag{5}$$

As the remnant expands, the energy of the relativistic particles varies as R^{-1}, and thus we have

$$L_v \propto K_0 \, (R_0/R)^{2\alpha}H^{\alpha}H_0 v^{-\alpha}. \tag{6}$$

The ratio H/H_0 depends on a variety of factors. In the early phases of the expansion it should be about 4 because the medium has no time to cool. Later, it becomes larger but how large depends on H_0, the interstellar particle density, and of course on the structure behind the shock front. When the shock slows down, the compression is again reduced and ultimately the field should become of uniform strength through the remnant. In these later phases, however, the relativistic electrons should also escape more easily and the remnant fades rapidly. Thus throughout the period in which the interstellar magnetic fields are of dominant importance, but before the final fading by escape of particles, we expect that H/H_0 does not change by a large factor (certainly not predictable!) and we treat it as a constant. For the remnants with $\log \Sigma$ less than 0, we then expect $L \propto R^{-0.8}$, which is in good agreement with the observed value. Obviously, our analysis is far from unique but at least it shows that a simple picture in accord with observations can be obtained.

During the very early phases of the remnant, the interstellar magnetic field can be of only limited importance because few field lines have yet been swept up and compression of more than a factor of four is not possible until cooling effects become important. However, because supernovae are almost certainly evolved stars, it appears likely that they will have lost some mass, which still could be in the neighborhood of the supernova. Thus there is no reason to suppose that circumstellar conditions are the same as those in the interstellar medium, with regard either to the density or to the magnetic field. This will not much affect the larger remnants, but in the early phases circumstellar conditions probably are dominant. Possible evidence for circumstellar matter and fields at the time of the explosion is provided by:

1) The stationary filaments in Cas A found by Baade and Minkowski. It is inconceivable that some filaments, initially moving at nearly 10,000 km/sec, could have been slowed down in such a short time without being completely dispersed. The only possible interpretation is that these filaments were pre-

sent before the supernova outburst. Possibly they were excited and ionized by the event. It is worth noting that an entirely comparable situation also is found in Tycho's and in Kepler's remnants, where the stronger filaments do not share the expansion velocity of the remnant.

2) The recent NRL polarization observations of Cas A have yielded the remarkable result that the average field in the synchrotron radiating region is radial rather than tangential. This indicates that the situation is not simply that the interstellar field is swept up by the shell. Possibly the field is stretched radially between the moving and stationary filaments.

3) Both the field structure and the distribution of field strength in the Crab Nebula are incompatible with the simple compression of interstellar magnetic fields.

4) Münch (1958) has pointed out that the major axis of the Crab Nebula is about parallel to the probable direction of the local interstellar magnetic field. It can easily be shown (Woltjer, 1958) that the interstellar field is completely inadequate to guide the present expansion of the nebula. However, the interstellar field could have significantly affected the motion of matter ejected before the supernova event, and have caused it to accumulate in such a way that the shell could less easily expand transverse to the original field direction.

Obviously, the importance of circumstellar conditions makes it very difficult to predict the behavior of young remnants and considerable differences between individual remnants can be expected. The situation is further complicated by the fact that the relativistic electrons cannot all have been injected at the time that the radius was very small (because $E \propto R^{-1}$). Thus, it appears likely that particle acceleration occurs for a considerable period (longer than 300 years in the Crab Nebula), and the time dependence of the particle production rate should influence the early evolution. The net effect will be to decrease the initial rate of decline compared to the prediction of the Shklovsky theory. Thus the discrepancy for Cas A is in the expected direction.

V   Time Scales and Energetics

Suppose a supernova remnant expands without important radiative losses. After expansion by a factor of two, relativistic particles have already transferred half of their energy to the shell. Thus, the shell energy can be expected to be generally larger than the relativistic particle energy. The energy of the expanding configuration is partly kinetic and partly thermal. After the initial phases the shell mostly consists of swept up interstellar matter. Writing the constant total energy ε_0 in the form

$$\varepsilon_0 = \eta \, (2\pi/3) \, \varrho_0 R^3 V^2 \tag{7}$$

with V the velocity of the shock separating the remnant from the undisturbed interstellar matter, we have from Sedov's (1959) similarity solutions $\eta = 1.37$ for $\gamma = 5/3$. Integrating, and taking $R = 0$ at $t = 0$, we obtain

$$R = (15/4\pi)^{2/5} \, [\varepsilon_0/(\eta\varrho_0)]^{1/5} \, t^{2/5} = (5/2) \, Vt. \tag{8}$$

In passing, we note that for the Crab Nebula, this equation is certainly not fulfilled because the Nebula has expanded without retardation. This again shows the complexity of the early phases.

As the shell expands and slows down radiative losses become important. An accurate evaluation of these losses is very difficult, mainly because of our lack of knowledge of inhomogeneities. We thus make a highly simplified picture. We suppose that for shell radii less than R_c all radiative losses can be neglected, but that for $R > R_c$ all material that is added to the shell radiates all its thermal energy. The value of R_c is determined from the condition that for $R = R_c$ the product of the expansion time R/V with the radiative loss rate just behind the shock, which separates the shell from the undisturbed interstellar matter, is equal to the thermal energy per unit volume behind the shock. At the time corresponding to $R = R_c$ there is of course gas inside the shell; but because this is hotter and much less dense than the gas near the shock (see for example the graphs by Heiles [1964] of the relevant solutions) the radiative losses will be small and this gas will gradually transfer most of its energy to the shell.

Behind the shock we have for the pressure P, electron density n_e, and temperature T

$$P = 3/4\varrho_0 V^2 = 3/4 n_0 m_0 V^2 \simeq 3/2 n_e kT = 6 n_0 kT. \tag{9}$$

The thermal energy per unit volume equals $3/2P$. The radiative loss rates have been calculated by a number of authors; the accuracy is still limited by poorly known cross-sections. From the computations by Pottasch (1965) we find that for the temperature interval $10^5 - 4 \times 10^6 \,^\circ K$—which is the most relevant for our discussion—the loss rate per unit volume can be represented, well within a factor of two, by $Qn_e^2 T^{-1}$, with $Q = 8 \times 10^{-17}$ c.g.s. The calculations by Tucker and Gould (1967), which overlap with those of Pottasch between 10^6 and 4×10^6 °K, give values on the average a factor of two smaller. Part of the difference is due to a somewhat different composition; also Pottasch did not include the effects of dielectronic recombination. Over most of the temperature range, however, the composition adopted by Pottasch does not lead to significant differences with results obtained for a more standard composition.

The radius R_c now is given by the condition that $3/2P$ be equal to $Qn_e^2 T^{-1}$ R/V or making use of equation (9)

$$\frac{V_c^5}{R_c} = \frac{1024}{9} \frac{k}{m_0^2} Qn_0. \tag{10}$$

During the late phases of the evolution of the shell when the internal energy is negligible and thus only kinetic momentum is conserved by the shell the velocity is inversely proportional to the mass and thus we will write

$$\varepsilon = \varepsilon_0 \frac{V}{V_c} \tag{11}$$

corresponding to

$$R^3 V = R_c^3 V_c. \tag{12}$$

We note that equation (12) is not very accurate because at the time corresponding to $R = R_c$ part of the total energy still is in thermal form and is transferred to the shell only somewhat later. Integrating equation (12) we have

$$R^4 = 4R_c^3 (V_c t - 3R_c/20) = 8R_c^4/5 (t/t_c - 3/8). \tag{13}$$

Eliminating R_c with equation (12) in terms of the radius and velocity at the time of observations R^* and V^* we write equation (10) as

$$\left(\frac{V_c}{V^*}\right)^{10/3} = 3.3 \times 10^{17} n_0 R^* V^{*-5}$$

$$= 1 \times 10^{11} n_0 (\text{cm}^{-3}) R^* (\text{pc}) V^{*-5} (\text{km/sec}). \tag{14}$$

If we again assume that at present the thermal energy is small compared to the kinetic energy we have

$$\varepsilon_0 = \varepsilon^* \frac{V_c}{V^*} = \frac{V_c}{V^*} \frac{2\pi}{3} R^{*3} \varrho_0 V^{*2}$$

$$= 1.2 \times 10^{42} \frac{V_c}{V^*} R^{*3} (\text{pc}) n_0 (\text{cm}^{-3}) V^{*2} (\text{km/sec}). \tag{15}$$

We note that if the thermal energy were not negligible the results would not change much because then the kinetic energy is less than follows from this equation. In fact, we have already seen in equation (7) that equating the total energy to $1/2MV^2$ does not introduce serious errors even in the early phases.

For the Cygnus Loop we make the very uncertain estimate $n_0 = 0.2$ cm^{-3} on the basis of Parker's (1967) photometry. With $R^* = 20$ pc and $V^* = 100$ km/sec we obtain $V_c = 200$ km/sec and $\varepsilon_0 = 4 \times 10^{49}$ ergs. For IC 443 assuming $n_0 = 1$ cm^{-2}, $R^* = 10$ pc and $V^* = 100$ km/sec we obtain $V_c = 240$ km/sec and $\varepsilon_0 = 2 \times 10^{49}$ ergs. Obviously the present analysis has been extremely rough and a quantittative study of the transition regime from the Sedov solution to the momentum conserving solution is much needed. It would be very surprising, however, if the present rough analysis caused errors of more than a factor of two in the values of ε_0.

For the younger remnants $\varepsilon^* = \varepsilon_0$. For the Crab Nebula with $M = 1 M_\odot$ and $V = 1400$ km/sec we have $\varepsilon_0 = 2 \times 10^{49}$ ergs. In a study of the filaments of Cas A Minkowski (1959) found densities of about 10,000 cm^{-3} and a total mass of about 1 $M_\odot$. The latter value was based on the assumption that the visible filaments are uniformly filled with matter and that only one tenth of all filaments are visible. We note that with a diameter of 0.02 pc and for an assumed temperature of 40,000 °K the H_α emission measure would be about 2×10^5. Allowing for a visual absorption of 5 magnitudes the observable emission measure would become 5000. Although the available data do not allow us to make quantitative estimates it is clear that most filaments and certainly those that are invisible have much lower emission measures. Thus the fraction of the volume of the filaments filled with matter is much less than unity and the mass of the filaments probably is smaller than that of the sun by one or two orders of magnitude. If we take $M = 0.1$ $M_\odot$ we would have $\varepsilon_0 = 5 \times 10^{49}$ ergs. We cannot exclude, however, the presence of a much larger mass of very hot gas and thus the total energy of Cas A remains essentially undetermined.

From equations (8) and (13) the ages of the remnants can be estimated. For the Cygnus Loop we obtain about 50,000 years, for IC 443 half as much. A general calibration of the $\Sigma - R$ curve in terms of age cannot be made because ϱ_0 and probably ε_0 differ for different remnants.

There are 14 remnants in Table I which are located in the galactic plane within 2 kpc from the sun. Following our earlier discussion we take 12 of these to be of Type II. This corresponds to 600 such remnants in the whole galaxy. Estimating very crudely that the average age is equal to the mean for IC 443 and the Cygnus Loop we arrive at a formation rate of one per sixty years. Obviously this value is very uncertain.

For the remnants in the halo the situation will be somewhat different. First of all, the interstellar magnetic field is likely to be smaller than in the disk. This may cause the radio luminosity to be less than that corresponding to our Σ-R relation. Further the interstellar density should be much smaller and thus the expansion velocity should be higher than for disk objects. Thus the time scales for the halo objects are shorter, which is compatible with the statistical evidence discussed earlier. The remnant of SN 1006 illustrates these points. If the object indeed followed the Σ-R relation, then its present size indicates a mean expansion velocity of approximately 20,000 km/sec. If the surface brightness is intrinsically less, however, the distance, and thus the velocity, become smaller and more reasonable. Nevertheless, it is clear that the object has undergone a fast expansion to large size, indicating that it has met little resistance.

VI   The Nature of Supernovae

The similarity between Type II supernova spectra and the spectra of ordinary novae has been noted repeatedly. The appearance of these spectra suggests that in the supernova process a shell is expelled which initially is optically thick. During the initial phase the spectrum is mainly a bluish continuum. Due to expansion the shell becomes optically thin and the continuum is replaced by emission lines.

The only Type II supernova for which rather complete UBV photometry is available is the supernova in NGC 7331 studied by Arp (1961). From a discussion of the colors at maximum—which are based on a somewhat uncertain correction for interstellar reddening—Arp found the effective photospheric radius at maximum light to be 6×10^{14} cm. Adjusting his values to a Hubble constant of 100 km/sec/Mpc we obtain a radius of 4.5×10^{14} cm and integrating under the light curve a total bolometric luminosity of 3×10^{49} ergs.

In the previous section we found that the mechanical (or thermal) energy injected by a typical Type II supernova is about 4×10^{49} ergs. At an estimated initial velocity of 10,000 km/sec this corresponds to 0.04 $M_\odot$ or at maximum to a surface density of 30 g/cm². This appears adequate to produce the black body-like continuum. Also one would expect that after some further expansion a shell of this mass would become transparent and produce an emission line spectrum in accordance with estimates made by one of us earlier (Poveda 1964).

Obviously, from the currently available data on supernovae, we cannot *exclude* the possibility that during the supernova outburst much more energy is released in the form of X rays or in other unobservable modes, but neither is there much evidence in support of this. Thus it would appear that there is little reason to believe that more than 10^{50} ergs is involved in a Type II supernova event. Such an energy could be produced by the burning of 0.007 $M_\odot$ of hydrogen into helium or 0.05 $M_\odot$ of helium into carbon.

Current theory connects supernovae of Type II with the ultimate collapse of rather massive stars in which an energy of 10^{53}–10^{54} ergs could be released. Even taking into account that a large part of this energy would be lost in neutrino and gravity wave modes, it seems that it is difficult to reconcile such a large energy release with the present state of the remnants. We note that the collapse-explosion theories of the supernova phenomenon can account for the supernova light only in a rather "ad hoc" fashion. The explosion leads to strong differential expansion, and thus, to cooling of the ejected matter. It is necessary to introduce post-explosion heating mechanisms to obtain adequate optical radiation (Colgate and White, 1966; Cameron, 1967). It has not been shown that in this way the evolution of the supernova spectra can be understood: certainly the hydrogen lines might be somewhat unexpected if the radiation arose from deep layers.

The analogy of the spectra of supernovae and novae suggests that similar mechanisms may be involved. In view of the fact that the total energy release need not exceed that corresponding to the helium burning of 0.05 $M_\odot$, it is tempting to think of instabilities in helium burning shells as the cause of the Type II supernovae. Rather weak instabilities of this type have been discussed by Härm and Schwarzschild (1965) and by Rose (1966). In the case of supernovae more violent instabilities would be required. Because of the difficulties of analyzing the Type I supernova spectra and because we do not have Type I remnants with adequate data, it is impossible to reach firm conclusions about them.

An observational test of the nature of the supernova phenomenon appears to be possible. If an instability is the cause, we would expect an observable stellar remnant to remain after the outburst. There is some ambiguous evidence that such an object is present in the Crab Nebula. A systematic search for stellar remnants would appear to be of much interest.

Even if our present ideas about supernovae are correct, we do not wish to imply that the collapse-explosion events do not take place. Certainly this mode of evolution leads to a plausible interpretation of cosmic rays and the synthesis of elements by processes of fast neutron capture. For cosmic-ray acceleration, one massive collapse supernova per 2000 years suffices (Colgate and White, 1966). Probably the same applies for nucleogenesis. Perhaps these events ultimately lead to objects of the kind discussed by Westerlund and Mathewson (1966). In any case, there is no reason to believe that they are related to the more common supernovae and supernova remnants.

This research was done during a visit of one of us to Columbia University, where it was supported in part by the National Aeronautics and Space Administration under Grant NsG 445, and in part by the Air Force Office of Scientific Research under Contract 49 (638) 1358, with Columbia University.

References

Arp, H. (1961) *Astrophys. J.* 133, 883.
Cameron, A.G.W. (1967) To be published.
Caswell, J.L. (1967) *Monthly Notices R.A.S.* 136, 11.
Chodil, G., Mark, H., Rodrigues, R., Seward, F., Swift, C.D., Hiltner, W.A., Wallerstein, F., Mannery, E.J. (1967) *Phys. Rev. Letters* 19, 681.
Clark, G.W., Garmire, G., Oda, M., Wada, M., Giacconi, R., Gursky, H., Waters, J. (1965) *Nature* 207.
Colgate, S.A. and White, R.H. (1966) *Astrophys. J.* 143, 626.
Conway, R.G., Kellerman, K.I., and Long, R.J. (1963) *Monthly Notices R.A.S.* 125, 261.
Friedman, H., Byram, E.T. and Chubb, T.A. (1967) *Science* 156, 374.
Härm, R. and Schwarzschild, M. (1965) *Astrophys. J.* 142, 855.
Heiles, C. (1964) *Astrophys. J.* 140, 470.
Kellerman, K.I. (1964) *Astrophys. J.* 140, 969.
Kraft, R.P. (1965) in *Stars and Stellar Systems*, Vol. V (ed. A.Blaauw and M.Schmidt, U. of Chicago Press) p. 162.
Mayer, C.H., McCullough, T.P., Sloanaker, R.M. and Haddock, F.T. (1964) *Astron. J.* 69, 552.
Menon, T.K. and Williams, D.R.W. (1966) *Astron. J.* 71, 392.
Minkowski, R. (1958) *Revs. Modern Phys.* 30, 1048.
– (1959) *Paris Symposium on Radio Astronomy* (ed. R.N.Bracewell, Stanford University Press) p.315.
Münch, G. (1958) *Revs. Modern Phys.* 30, 1042.
Parker, R.A.R. (1967) *Astrophys. J.* 149, 363.
Pottasch, S.R. (1965) *Bull. Astron. Inst. Neth.* 18, 7.

Poveda, A. (1964) *Ann. d'Astrophys.* 27, 522.

Rose, W. (1966) *Astrophys. J.* 146, 838.

Sedov, L.I. (1959) *Similarity and Dimensional Methods in Mechanics* (New York: Academic Press).

Sharpless, S. (1955) *Astron. J.* 60, 178.

– (1959) *Astrophys. J.* 4, 257.

– (1965) in *Stars and Stellar Systems*, Vol. V. (Ed. A. Blaauw and M. Schmidt, U. of Chicago Press) p. 136.

Shklovsky, I.S. (1960) *Soviet Astron. A. J.* 4, 243.

Tucker, W.H. and Gould, R.J. (1966) *Astrophys. J.* 144, 244.

Westerlund, B.E. and Mathewson, D.S. (1966) *Monthly Notices R. A. S.* 131, 371.

Woltjer, L. (1964) *Astrophys. J.* 140, 1309.

– (1958) *Bull. Astron. Inst. Neth.* 14, 39.

P. Morrison   It seems to me that in considering the production of these enormous energies one should take into account the efficiencies of the various processes. For example, the efficiency for the production of relativistic electrons is only 10%.

L. Woltjer   In that case, one could say that the inefficiency of 90% comes out as x-rays, so that if we add up the relativistic electrons and the x-rays we end up with the same result. Therefore one doesn't have to worry about the efficiency; the details of the efficiency will depend upon the model that is used.

P. Morrison   It's not reasonable to say that the whole energy source is equal to only the energy that we see radiated. This is only a lower limit. There could be a great deal more.

L. Woltjer   In what form would it be stored?

P. Morrison   In hot plasma.

L. Woltjer   But in the end, isn't the hot plasma going to expand into the interstellar medium and add energy to the shell? In an object like the Crab Nebula, you have almost unlimited possibilities for storing energy. But this is not the case for the Cygnus Loop, so that as an upper limit it is a far more sensitive case.

P. Morrison   Then the historical supernovae are not part of the class. You are proposing another class of explosions which make these calmer objects. Have you studied Tycho's supernova from this point of view?

L. Woltjer   I have no idea what energy is involved. There is some x-ray and radio emission, and very little optically. All one can say is that it is developing in the constant energy phase.

S. Colgate   I would like to take issue with your comment on cosmic-ray cooling. It depends upon formation of a diamagnetic sphere, with charge separation providing the primary boundary for the diamagnetic effect. Once the energy exceeds a Bev/nucleon the electron mass is comparable to the ion mass, and we find that the cosmic-ray particles will escape the sphere very rapidly.

L. Woltjer   If that kind of model is used, then one can say that afterwards a new set of cosmic rays is produced, and that is what is seen in the remnant. What I wished to say was that the cosmic-ray particles seen in supernova remnants are of no use in explaining galactic cosmic rays because they leak out after the remnant has expanded.

4

COMPUTATION OF RELATIVISTIC GRAVITATIONAL COLLAPSE

R. A. Schwartz

Brown University
Providence, R. I.

THE AIM of this paper is to discuss the collapse of stars as predicted by general relativistic calculations. In particular, we are attempting to answer the question as to the ultimate fate of a star having a mass greater than the Chandrasekhar limit or the neutron star limit. Will the star explode as a supernova, shedding enough mass to trim down below the appropriate mass limit? Will it collapse indefinitely into a singularity? Will, in fact, both processes occur? All of these possibilities are possible in principle, but we must perform the numerical calculations in order to determine the conditions under which the various possibilities will occur.

We will begin with an outline of the method of calculation. The metric for a spherically symmetric star is

$$ds^2 = -e^{-2\varphi}\,dt^2 + \frac{1}{\gamma^2}\,dR^2 + R^2\,d\Omega^2 \tag{1}$$

Since the radius is not a well-defined idea, R actually represents a circumference; i.e., the circumference of a circle in this system is $2\pi R$. We propose to solve the Einstein equation

$$G_{\mu\nu} = 8\pi T_{\mu\nu} \tag{2}$$

where $G_{\mu\nu}$ is the divergenceless Ricci tensor and $T_{\mu\nu}$ is the stress-energy tensor. We have adopted a set of units in which $c = G = 1$. We write the stress-energy tensor in the form

$$T_{\mu\nu} = t_{\mu\nu} + \Delta T_{\mu\nu}. \tag{3}$$

Here $t_{\mu\nu}$ represents the stress-energy tensor for a perfect fluid, as follows:

$$t_{\mu\nu} = \begin{pmatrix} \varrho\,(1+\varepsilon) & 0 & 0 & 0 \\ 0 & P & 0 & 0 \\ 0 & 0 & P & 0 \\ 0 & 0 & 0 & P \end{pmatrix} \tag{4}$$

where ϱ is the proper density of particles, ε is the total internal energy, and P is the scalar pressure. The anisotropic portion of the stress-energy tensor is given by

$$\Delta T_{\mu\nu} = \begin{pmatrix} \varrho\,\Delta\varepsilon & & \vec{q} & \\ & \Delta P_{\rm II} & 0 & 0 \\ \vec{q} & 0 & P_{\perp} & 0 \\ 0 & 0 & P_{\perp} \end{pmatrix} \tag{5}$$

In this case $\Delta\varepsilon$ represents the energy density in the anisotropic part of the radiation field (in particular, the energy carried by neutrinos). $\vec{q}$ represents the energy flux, and $P_{\perp}$ and $P_{\rm II}$ represent the components of the pressure due to the anisotropic radiation field.

The equations of hydrodynamics must be written in relativistic form. The equation of continuity becomes

$$(\varrho u^{\mu})_{;\mu} = 0 \tag{6}$$

The analogue of the Euler equations of hydrodynamics becomes

$$T^{\mu\nu}{}_{;\nu} = 0 \tag{7}$$

If we multiply by the velocity u_{μ}, we obtain a relativistic analogue of the first law of thermodynamics:

$$u_{\mu}(T^{\mu\nu}{}_{;\nu}) = 0 \tag{8}$$

We will now proceed to convert these equations into a form such that they can be solved numerically. We define the co-moving proper time derivative $D_t = e^{-\varphi}(\partial/\partial t)$. Then we can define a velocity $u = D_t R$. We perform a change of variables to a coordinate μ. We thus change the metric by substituting $dR^2 = (dR/d\mu)\,d\mu^2$. We then obtain from the Einstein equation (Eq. 2) the following:

$$D_t u = \frac{\gamma}{W}\,4\pi R^2 \varrho P' - \frac{(m + 4\pi R^3 P/3)}{R^2} \tag{9}$$

A prime (′) after a quantity is defined to mean $\partial/\partial\mu$ of that quantity. In the above, $\gamma = |4\pi\varrho R^2 R'|$ and $W = 1 + \varepsilon + P/\varrho$. The mass m is defined by the equation

$$m(\mu) = \int_0^\mu (1 + \varepsilon)\, 4\pi\varrho R^2\, dR$$

The ε takes into account the fact that the energy density is an additional source of the gravitational field.

The equation of continuity (Eq. 6) becomes

$$D_t (\ln \alpha) = \frac{u'}{R'} - \frac{L}{R} \tag{10}$$

where $\alpha = \varrho R^2$ and L is the luminosity defined as $4\pi R^2 q$. Finally, Eq. 8 reduces to

$$D_t \varepsilon = -PD_t \left(\frac{1}{\varrho}\right) - e^{-2\varphi} (L\, e^{2\varphi})' - Q_L \tag{11}$$

where Q_L represents energy losses due to effects such as mu neutrino emission. These equations have been obtained by several authors (cf. Misner and Sharp, 1964 or Schwartz, 1967).

The equations cannot be solved exactly, so we must transform them into finite difference equations to be solved numerically. The star is divided up into shells (see Fig. 1); we compute the radius R_j^n and the velocity u_j^n at the jth shell, the $j + 1$st shell, etc., at time step n. For the gas between these shells we determine the thermodynamic quantities such as the density $\varrho_{j+1/2}^n$ and temperature $T_{j+1/2}^n$. As boundary conditions we have that at the center of the star $R = 0$ and $u = 0$; at the surface of the star $\varrho = 0$ and $P = 0$. With this scheme we can compute the values of all quantities of interest in each zone for successive time steps.

The first step in the calculation is to advance the momentum equation (Eq. 9) in time using the finite difference scheme

$$\frac{u_j^{n+1/2} - u_j^{n-1/2}}{\Delta t^n} = a_j^n f_j^n \tag{12}$$

where $a = e^\varphi$ and f_j^n denotes the right-hand side of Eq. 9, with all the quantities appropriately centered in space at zone j and in time at step n. Now that we have the new velocities, we can calculate the positions of the zone boundaries by

$$\frac{R_j^{n+1} - R_j^n}{\Delta t^{n+1/2}} = \bar{a}_j^{n+1/2} u_j^{n+1/2} \tag{13}$$

Here we rely on the fact that the function a is slowly varying, so that we may extrapolate forward in time:

$$\bar{a}_j^{n+1/2} = a_j^n + \frac{1}{2}\left(\frac{\Delta t^n}{\Delta t^{n-1}}\right)(a_j^n - a_j^{n-1}) \tag{14}$$

The next step is to advance the density. From Eq. 10 we have

$$\frac{(\Delta \ln \alpha)_{j+1/2}^{n+1/2}}{\tilde{a}_{j+1/2}^{n+1/2}\Delta t^n} = \frac{u_{j+1}^{n+1/2} - u_j^{n+1/2}}{\bar{R}_{j+1}^{n+1/2} - \bar{R}_j^{n+1/2}} - \frac{\tilde{L}_{j+1/2}^{n+1}}{\tilde{R}_{j+1/2}^{n+1/2}} \tag{15}$$

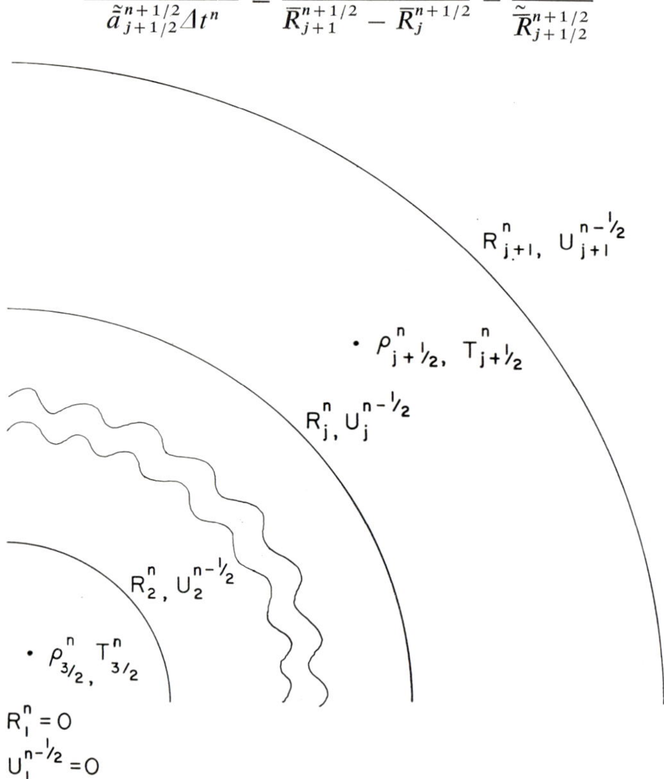

Fig. 1 Division of star into zones for numerical computations. The mechanical quantities R and u are associated with zone boundaries; the thermodynamic quantities ϱ, P, ε and T are associated with zone centers.

In the above, a bar $(-)$ over a variable means that it has been extrapolated forward in time; a tilde $(\sim)$ means that it has been extrapolated forward in space. We now apply the energy equation

$$(\varDelta T)^{n+1/2}_{j+1/2} = \frac{\left[P + Q + \dfrac{\partial \varepsilon}{\partial (1/\varrho)}\right]^{n+1/2}_{j+1/2} \left[\dfrac{1}{\varrho^{n+1/2}_{j+1/2}} - \dfrac{1}{\varrho^{n}_{j+1/2}}\right]}{\left(\dfrac{\partial \varepsilon}{\partial T}\right)^{n+1/2}_{j+1/2}}$$

$$- \left\{ L^{n+1}_{j+1} - L^{n+1}_{j} + 2\, \tilde{L}^{n+1}_{j+1/2}\, \frac{P^{n+1}_{j+1/2} - P^{n+1}_{j-1/2}}{\widetilde{W}^{n+1}_{j}\, \tilde{\varrho}^{n+1}_{j}\, \varDelta \mu_j} \right\} \tag{16}$$

where Q is the Von Neumann-Richtmyer artificial viscosity,

$$Q = \begin{cases} \tfrac{1}{2}\varrho\,(du)^2, & \varDelta\varrho > 0 \\ 0, & \varDelta\varrho \le 0 \end{cases}$$

We compute the neutrino luminosity

$$L^{n}_{j} = -\,\frac{14\pi^2 a}{3K}\,(R^{n}_{j})^4 \left\{ (T^{n}_{j+1/2})^4 \right.$$

$$\left. -\,(T^{n}_{j-1/2})^4 + (\tilde{T}^{n}_{j})^4 \cdot 4\,\frac{P^{n}_{j+1/2} - P^{n}_{j-1/2}}{\widetilde{W}^{n}_{j}\, \tilde{P}^{n}_{j}\, \varDelta \mu_j} \right\} \tag{17}$$

The opacity K used here is an appropriate analogue of the Rosseland mean, as derived in Schwartz (1967). The stability conditions for these difference equations are also discussed in this paper.

The procedure is followed for each zone, starting from the center outward. Since the boundary condition is imposed at the outer boundary of the star to make t coincide with the time measured by an observer on that boundary, we must now integrate inward to obtain

$$\varphi^{n+1}_{j} = \varphi^{n+1}_{j+1} - \frac{\tilde{P}^{n+1}_{j+1} - \tilde{P}^{n+1}_{j}}{W^{n+1}_{j+1/2}\, \varrho^{n+1}_{j+1/2}} \tag{18}$$

starting with $\varphi_{j\,\mathrm{max}} = 0$. Then we may calculate γ^{n+1}_{j+1} and m^{n+1}_{j+1} in a straightforward way, and we are ready to go on to the next time step.

The results of these calculations in the case of continued gravitational collapse are shown qualitatively in Fig. 2. As time progresses, the geometry becomes increasingly curved until the star pinches itself off from the world.

The star will gradually disappear as the photons get more and more red-shifted. This is a real singularity which can't be avoided. Some stars will, of course, explode as supernovae, but it is not necessarily the case that every star that ends its energy-generating lifetime with a mass above the Chandrasekhar limit or the neutron star limit will explode as a supernova. It can collapse, dissipating its energy in some undetectable form such as in neutrinos, and go directly to the singularity without ever exploding. Quantitative results of the calculations may be found in Schwartz (1967, 1968).

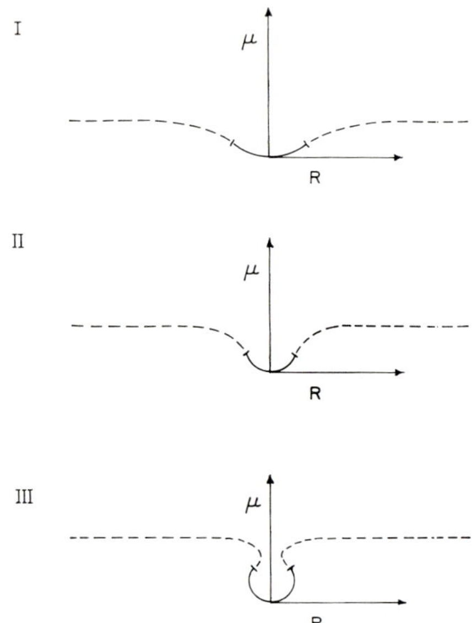

Fig. 2   Change in geometry as collapse proceeds.

References

Misner, C.W. and D.Sharp, Phys. Rev. **136**, B571, 1964.
Schwartz, R.A., Ann. Phys. **43**, 42, 1967
Schwartz, R.A., To be published (submitted to Ann. Phys.), 1968.

DISCUSSION

R. Macklin   What sort of equation of state did you use for the matter in the star?

R. A. Schwartz   I assumed an ideal Fermi gas for nuclear matter. Other equations of state could be used which would make quantitative but not qualitative changes in the solution. What is important is the behavior of the star with respect to the neutrinos it emits; i.e., at what point does the star become optically thick to neutrinos? This is the key to whether or not the star will explode as a supernova.

P. Thaddeus   What can you say about the stability of these models? For example, what happens if the star is disturbed with some non-radial perturbation?

R. White   If the star is far enough into the collapse, it will tend to remain spherical. The quantity γ which appears in Eq. 9 will have the value $\left(I + u^2/c^2 - \dfrac{2GM}{Rc^2} \right)^{1/2}$ when the star is in continued collapse. Consequently γ will approach zero and the pressure gradient will have no effect. The tendency is then to symmetrize the solution.

R. A. Schwartz: The stability to nonradial perturbations of an object in continued gravitational collapse is likely to be the same as that of the Schwarzschild solution itself, since, as White pointed out, the pressure gradient has only a small effect. An analysis of the stability of the Schwarzschild solution has recently been given by Vishveshwara (thesis, U. of Maryland, 1968) using a variation of the Regge-Wheeler method of tensor spherical harmonics to treat perturbations of the Kruskal form of the Schwarzschild metric. He finds, not surprisingly, that the solution is stable to small non-radial perturbations.

5

SUPERNOVA HYDRODYNAMICS

S. A. Colgate

New Mexico Institute of Mining and Technology
Socorro, New Mexico

IN THIS PAPER, I would like to discuss two rather different aspects of the supernova phenomenon. The first of these concerns the implications with respect to energy requirements of the early light curves of supernovae. The second concerns dynamics of the shock wave from the standpoint of neutrino diffusion. In particular, it is a question of whether or not the neutrino transport in the shock wave will be enough to enable the star to avoid the wormhole catastrophe. The details of this question represent a point of disagreement between the results obtained by Arnett and by White and me, but it should be noted that this is a relatively minor departure in view of the considerable agreement between our respective results.

In analyzing the light curves of the most energetic supernovae, we would like to be able to determine whether the energy needed to produce a supernova can be supplied entirely by nuclear binding energy or whether gravitational binding energy must be included. The brightest supernovae are approximately 25 magnitudes brighter at maximum than the sun. This corresponds to a luminosity 10^{10} times that of the sun, or about 4×10^{43} erg/sec. The light curve will typically have a half-width of about one week, or approximately 10^6 sec (Fig. 1). Consequently, the optical energy associated with the light curve maximum is about 4×10^{49} ergs. If the supernova surface were at the same temperature as the sun, then the luminosity would depend on R^2, so that the radius of the supernova at maximum would have to be 10^5 times the solar radius, or 7×10^{15} cm. On the other hand, if the supernova temperature were 30,000 °K the optical output would increase by a factor of 5 or 6, compared to a surface at 6000 °K (this is a consequence of the shapes of the Planck curves for these temperatures), but the bolometric correction would be very large. The total radiated energy would then be 10^{52} ergs, which we consider unrealistic. Therefore, it is more reasonable to assume a temperature of about 10,000°K. For this temperature, the radius will be about 3×10^{15} cm; if we assume that the supernova reaches maximum in 10^6 sec, it follows that the velocity of the surface is 3×10^9 cm/sec, which corresponds to about 5 Mev/nucleon.

We can make a rough calculation based on diffusion theory arguments of the total mass involved. The density of the expanding material is such that free-free transitions are negligible and the only opacity is provided by Compton scattering. If we assume that no energy is being generated at the time of maximum, that is, that the energy source has been turned off very early, then diffusion theory predicts that the maximum will occur at approximately the time when a diffusion wave has penetrated to about one third the radius of the sphere. Therefore, the diffusion velocity should be one third of the expansion velocity of the surface:

$$v_{\text{diff}} = \tfrac{1}{3} u_s. \tag{1}$$

From diffusion theory, the skin depth ∂ is given by $\partial^2 = Dt$, where D is the diffusivity. We substitute $D = \lambda c/3$, so that

$$\frac{\partial}{\lambda} = \frac{t}{\partial} \cdot \frac{c}{3}. \tag{2}$$

If we consider ∂/t to be the diffusion velocity v_{diff}, then from Eqs. (1) and (2) we obtain $\partial/\lambda = c/u_s = 10$ (since we previously calculated u_s to be 3×10^9 cm/sec). We interpret this result to mean that the medium is 10 mean free paths thick. This then requires that $\varrho r/3 = 10/K$, where K is the Compton opacity; taking $1/K = 5$ g/cm^3 gives $\varrho r = 150$ g/cm^2. A computer calculation using a more complicated diffusion theory and a variable velocity distribution gives a value of $\varrho r = 50$ g/cm^2. The total mass will then be $4\pi r^2 (\varrho r)/3$; using the more conservative value of $\varrho r = 50$ gives a mass of 2×10^{33} gms, which is one solar mass.

We make the further assumption that in an explosion of this magnitude the energy generation must occur in a time of the order of magnitude of the traversal time of a sound wave through the reacting region. Since the reacting region is held together by gravitation, it will follow that half the work will be done against gravity and the other half will go into kinetic energy. Therefore, the total energy generated is about 10 Mev/nucleon for a solar mass of ma-

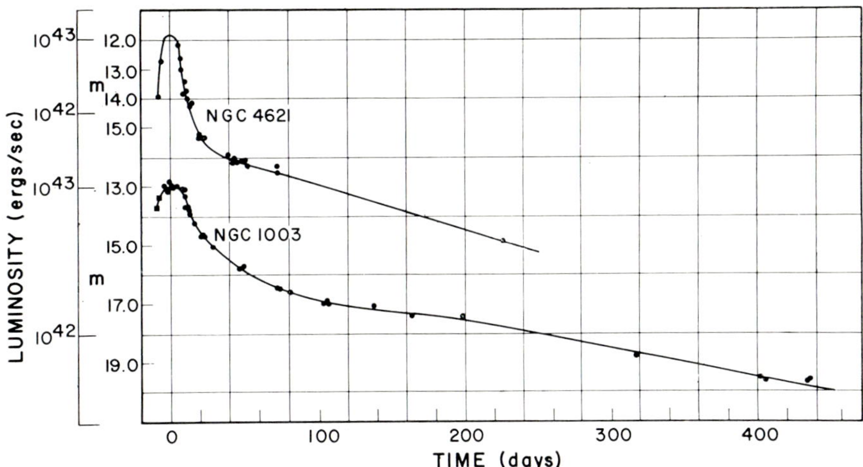

Fig. 1 Two typical supernova light-curves are shown giving a peak luminosity of 10^{43} ergs/sec at a time ≥ 5 days (courtesy Minkowski 1964).

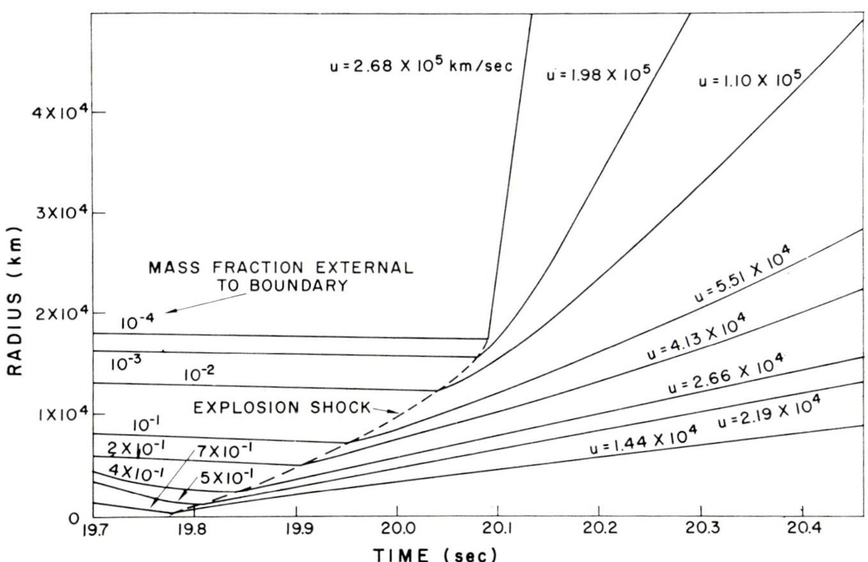

Fig. 2 10 $M_\odot$ supernova radius versus time. The linear plot shows the increasing velocity of the outward explosion shock wave reaching the relativistic limit for 10^{-4} mass fraction.

terial. This is almost an order of magnitude more energy than we can hope to get from nuclear reactions alone and indicates that gravitational binding energy must be utilized.

To understand the energetics of the explosion, we have considered the hydrodynamics in more detail. We have found, for example, that when the energy is released in a time that is short compared to the sound speed traversal time, the velocity distribution will be such that the outer layers of the star will expand at much higher velocities than the deeper layers; that is, the shock wave will speed up in a density gradient. This result is shown graphically in Fig. 2. The velocity distribution will be independent of the stellar structure, provided that velocity is plotted a as function of the external mass fraction F, defined as the ratio of the mass external to the point in question to the total mass of the system. The velocity distribution is calculated on the basis of the

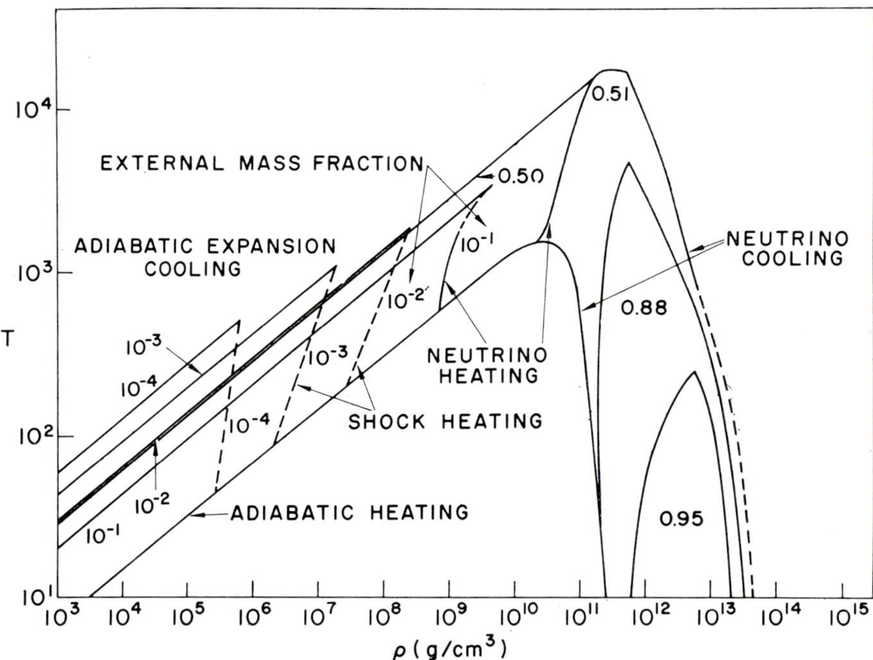

Fig. 3   10 M$_\odot$ supernova temperature versus density with neutrino deposition. The lower line of slope 4/3 corresponds to the initial adiabatic compression during implosion. The innermost zones cool by neutrino emission; the intermediate zones heat by neutrino deposition; and the outermost zones are shock heated. The expanding matter then cools adiabatically.

effects of the shock wave, as represented in Fig. 3. Following the implosion onto a neutron core, the neutrino heating of the inner regions causes an expansion wave that successively shocks the various outer zones of the star; we have assumed that these zones expand adiabatically.

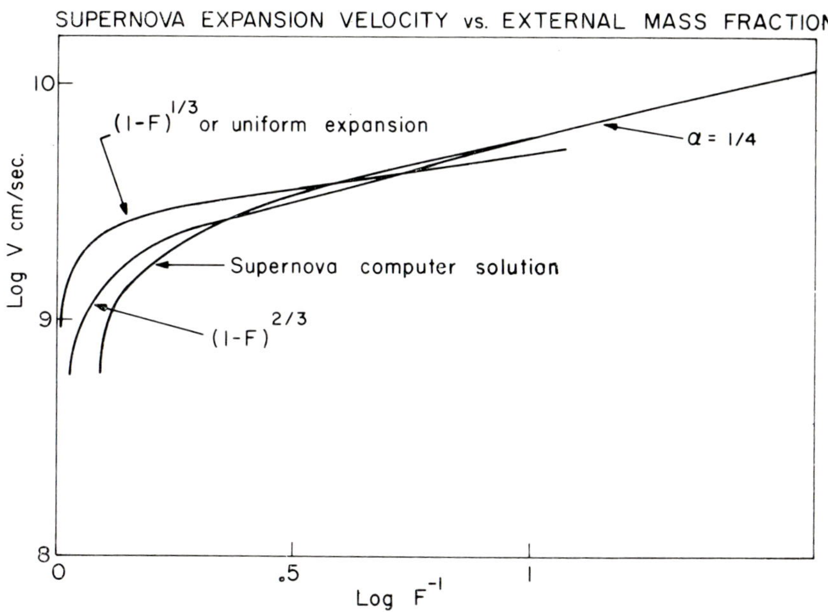

Fig. 4 The supernova ejection velocity is shown versus external mass fraction. Two analytic curves are drawn that best fit the computer solution.

The velocity distribution as a function of external mass fraction for a plane parallel mass distribution is shown in Fig. 4. The asymptotic behavior of the solution on a log-log plot has a slope $\alpha = 1/6$. This analytic solution is compared with the computer solution for a sphere of polytropic index 3. The computer solution exhibits similar behavior, with $\alpha = 1/4$; that is, the velocity distribution is given by

$$v = v_0 F^{-1/4} \tag{3}$$

for mass fractions up to about 30%. Our treatment is nonrelativistic, so this formula will not apply for mass fractions less than about 10^{-3}. For external mass fractions greater than about 30%, the velocity distribution varies approximately as $(1 - F)^{2/3}$, which is considerably slower than a uniform spherical expansion of the medium. The significance of this result with respect to

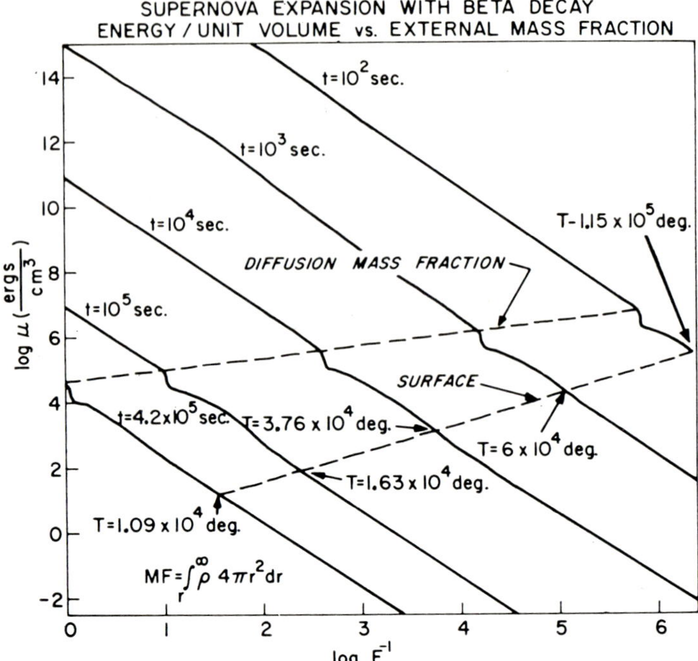

Fig. 5   The internal energy of the expanding matter is shown versus mass fraction at various times. These are analytic solutions that include radiation diffusion and beta decay energy. The outer dashed curve is the optical surface and the inner one corresponds to the mass fraction from which the energy is diffusing.

the energy source is that it predicts that a significant fraction of the mass expands away very rapidly; it will be essentially transparent, so that we tend to see deeper into the star to the slower-moving matter. This effectively doubles the energy requirement as predicted on the basis of the light curve alone.

Chester McKee and I have been able to get an analytic solution to the radiation flow problem by assuming the velocity distribution of Eq. (3). The results are shown in Fig. 5. The radiation is diffused from essentially the region between the two dashed lines as indicated by the small bump in the curve that progresses with time toward a larger mass fraction or toward the center of the star. The surface temperature can also be calculated analytically; the values obtained are of the order of $10^4\,°K$, which is quite reasonable. The calculation does not apply for times earlier than 10^3 sec, because the velocities will be relativistic; the mass fractions involved will be accelerated to energies greater than a Bev/nucleon, and the diffusion theory will not apply.

The most surprising result of our calculation is that the luminosity does not increase with time so rapidly as one might expect (Fig. 6). This comes about because the material through which the shock wave proceeds is opaque; the inner layers will have higher temperatures, but they will not be seen until after they have expanded adiabatically to a lower temperature. We are, in effect, averaging over the outermost layers, which leaves us with a slow increase in the luminosity with time. This result presents the possibility that if we observe a supernova early enough, we may see the surface layers corresponding to small mass fractions. A knowledge of the velocity distribution of the expanding matter is essential as a key to understanding the energetics of supernovae.

Now let us proceed to our second problem, namely, the transport of energy by neutrinos in the shock wave. The history of the collapse is depicted

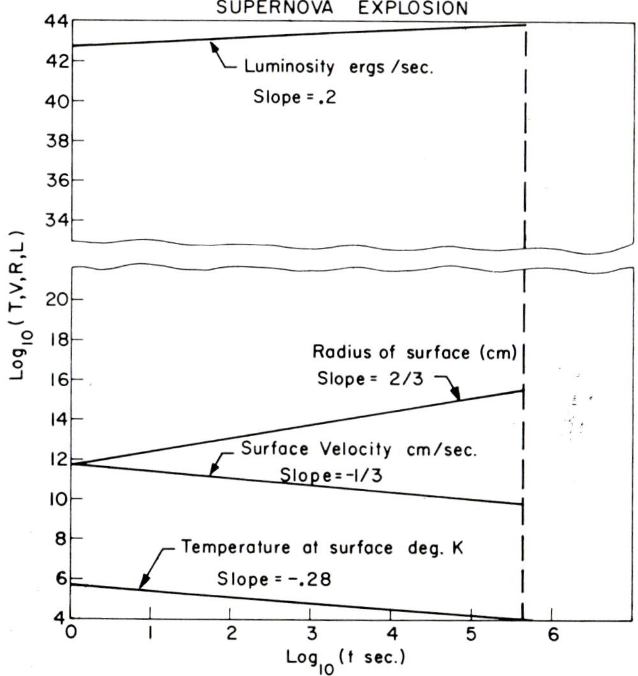

Fig. 6 The resulting temperature, radius, bolometric luminosity, and surface velocity for the analytic solution is shown versus time.

in Fig. 7. The question under consideration is whether the energy generated
is held by the in-falling matter and subsequently emitted slowly as mu-meson
neutrinos or whether it is able to diffuse out more rapidly in the form of an
electron neutrino flux causing heating and explosion. One difference between
our calculations and those of Arnett has to do with the equation of state.

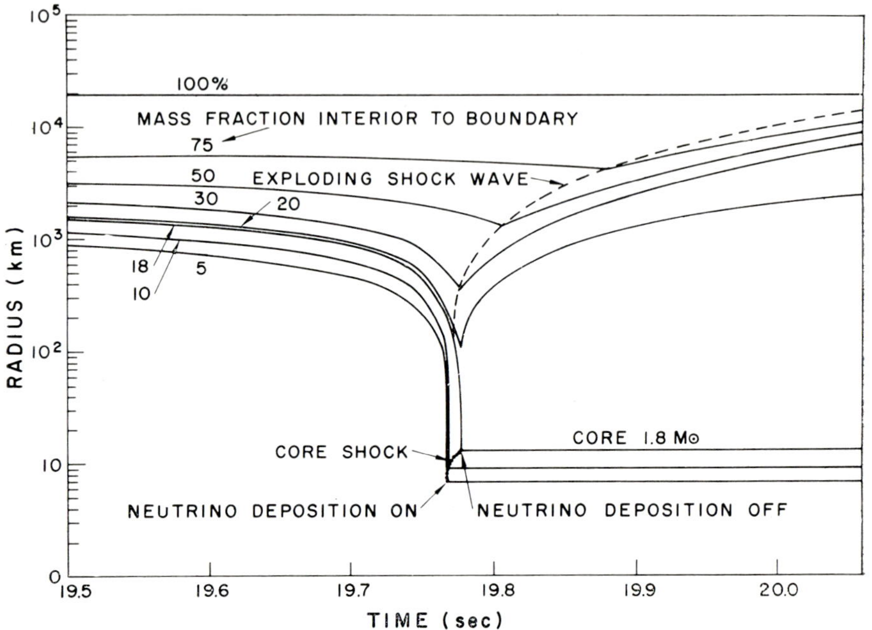

Fig. 7   Radius versus time for 10 M$_\odot$ supernova with neutrino deposition. During the initial collapse the neutrino energy is assumed lost from the
star, but at the time of formation of a core shock wave (heavy dots) a fraction of the neutrino energy is deposited in the envelope. The deposition
ceases when the explosion terminates the imploding shock wave on the core.

The deviation of our equation from that of Arnett, based on the cold Skyrme
potential, can be seen in Fig. 8. The pressure required to maintain equilibrium
in stars of 1.1 and 1.5 solar masses is also shown; the fact that the actual
pressure is almost an order of magnitude less than that required for equilibrium means that the collapse will take place very nearly in free fall. In this
respect, our results are quite similar; in calculating the time for a solar mass
to fall onto the neutron core, we obtain a value of 3 msec, while Arnett gets
5 msec.

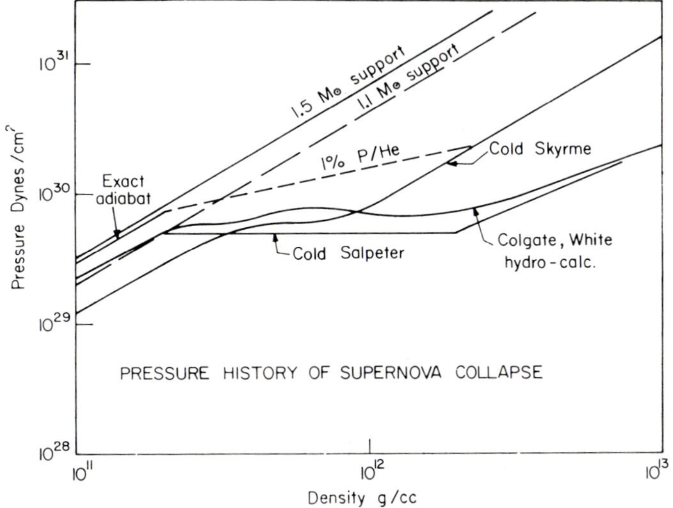

Fig. 8 Shows pressure versus density for cold matter using the Salpeter equation of state, the cold Skyrme equation of state, the pressure for the Colgate and White hydro-calculations, and the more correct pressure. 1% P/He, the Arnett calculation and the equilibrium pressure support required for a 1.1 $M_\odot$ and 1.5 $M_\odot$, respectively. The large pressure defect for equilibrium support for all equations of state is the reason for dynamical collapse.

The main difficulty, however, is the question of the electron neutrino diffusion. Figures 9 and 10 show the results of some arbitrary calculations meant to test the ability of our diffusion hydrodynamic code to reproduce a shock wave with diffusion processes. Each zone represents a mean free path. We previously calculated (Eqs. 1 and 2) that the skin depth (the width of the shock) is to the mean free path as $c/3$ is to the shock velocity. Since we arbitrarily chose a shock velocity of 10^9 cm/sec, we would expect the shock to be about 10 zones wide if we have represented the radiation flow properly; this is approximately the result we obtain. We then expanded the calculation to include many more zones (150 in this instance). We use an opacity proportional to T^2, which roughly represents the way in which neutrinos will behave. We also assign a specific heat to the gas; in the previous calculations all the energy behind the shock was in the form of radiation, but in this calculation some of the energy is stored as particle internal energy.

The significant conclusions of our calculation are as follows: We find that the temperature behind the shock is lower by a factor of 2 than in the previous

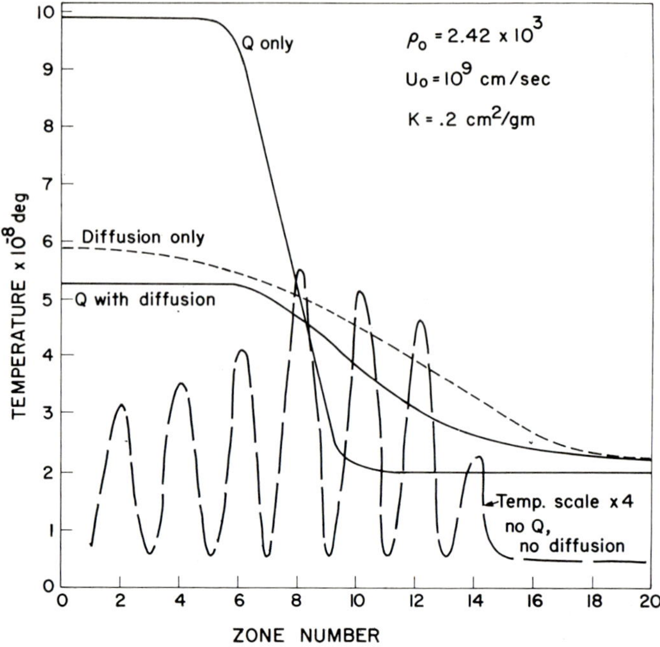

Fig. 9 The temperature versus zone number is shown for shocks produced (1) without artificial viscosity (Q) dashed curve, (2) with Q alone top curve, (3) with both Q and diffusion middle solid curve, and finally (4) with diffusion alone dotted curve.

$$\varrho_0 = 2.42 \times 10^3 \text{ g/cc}$$
$$U_0 = 10^9 \text{ cm/sec}$$
$$K = 0.2 \text{ cm}^2/\text{gm}$$
$$K\varrho_0 \, \Delta X = 1 \text{ mean free path}$$
$$\tau = 1.32 \times 10^{-11} \text{ sec.}$$

solutions where all the energy is in radiation. The internal energy density in this region is very low (Fig. 9), which indicates that the region has been very highly compressed and is highly opaque to radiation; that is, the energy generated in the shock will not radiate backward and so will not heat the material behind the shock. The mean free path behind the shock is very small; about 95% of the neutrino energy is radiated ahead of the shock. One zone in Arnett's calculations corresponds to a thousand of ours. His single zone will reach a higher temperature and will be able to radiate internal energy by emitting mu-neutrinos. In our calculations, the energy will not be radiated in this form because the temperature will not be so high. Thus, we feel that the star is more likely to explode. In particular, we find that the more massive stars, with more than 4 solar masses in the core, will explode, whereas Arnett finds that they will not.

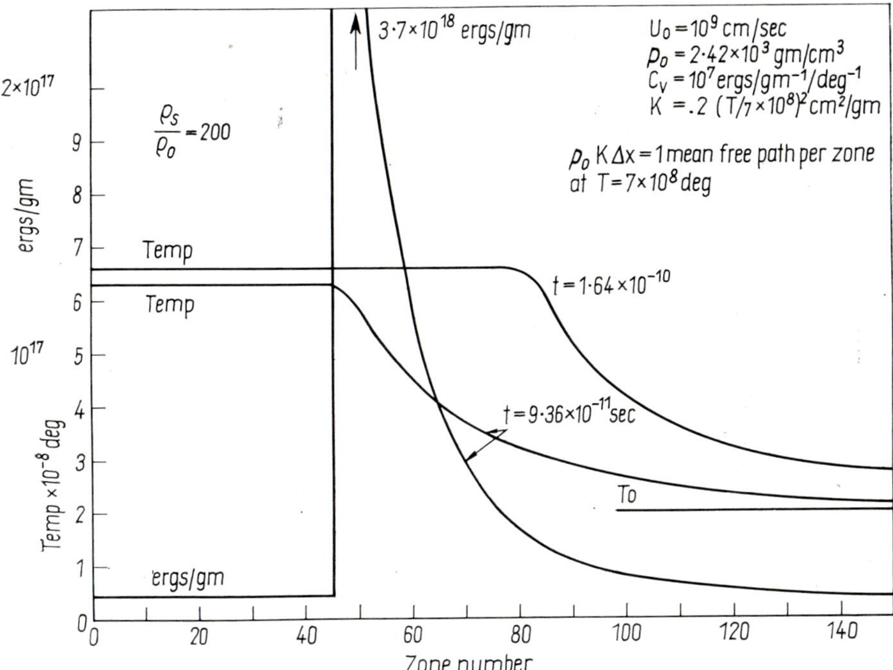

Fig. 10 Temperature and internal energy density versus zone number for a shock produced with an opacity, $K \sim T^2$. A precursor of radiation extends well ahead of the shock and as a consequence, the temperature is slow to reach the asymptotic value, and the internal energy per gram is held way down despite a spike in front. The shock has traversed a distance (100 zones) large compared to its minimum thickness (10 zones) and shows a major departure from the asymptotic condition $\varepsilon_0 = 5 \times 10^{18}$ ergs/gm., $U_0 = 10^9$ cm/sec, $\varrho_0 = 2.42 \times 10^3$ gm/cc, $C_0 = 10^7$ ergs/gm deg., $K = 0.2 \left(\dfrac{T}{7 \times 10^8} \right)^2$, $\varrho_0 K \Delta X = 1$ mean free path at $T = 7 \times 10^8$ deg.

DISCUSSION

R.Schwartz   What is the basis for your choice of opacity formula?

S.Colgate   Compton opacity because of low density. There is also a question of whether one should use diffusion theory or transport theory. A basic criterion for using diffusion theory is that the mean free path must not be long compared to the dimension of the logarithmic temperature gradient. In this distance, the mean free path was about one tenth of the gradient, so the use of the diffusion theory was reasonable. In general, diffusion theory gives a more conservative estimate of the energy transfer than transport theory.

J.Wheeler   When I discussed your calculations with Fowler, he brought up a point which I don't know how to assess. Namely, he feels that as the outer layers fall in, nuclear reactions will develop before the density is anywhere near what is required to give a significant neutrino opacity, and the envelope will be driven off.

S.Colgate   This was, in fact, our original concept based upon Burbidge, Burbidge, Fowler, and Hoyle as to how a supernova should explode.

However, we found that releasing 2–4 MeV/nucleon from thermonuclear reactions would not prevent the star from continued collapse. Even with the most extreme assumption for thermonuclear energy release, the outer layers went out only a little and then fell back in. As a result of neutrino heating, the energy in the shock wave is about 50 MeV/nucleon which will make the star explode.

A.Poveda   How much energy goes into relativistic particles?

S.Colgate   About 10^{50} ergs. However, if the envelope has a red giant structure, there would be a slightly smaller mass fraction and less energy in cosmic rays but more visible light.

G.Field   Where does the energy of the most energetic supernovae finally reside?

S.Colgate   I can only guess at that. It is interesting, though, that some calculations have been made (Spitzer) in which the heating of HI regions can be accounted for by extrapolating the cosmic ray spectrum down to 5–10 Mev/nucleon, which would represent the major fraction of the energy of the supernova.

6

EXPLODING STAR MODELS
AND SUPERNOVAE

W. David Arnett

California Institute of Technology
Pasadena, California

Abstract

Although supernova remnants are observed in other frequency ranges, supernovae are an optical phenomenon. Numerous attempts have been made to relate observations of these events to explosive instabilities in the theory of stellar evolution. This paper will discuss some aspects of the theory of the late stages of stellar evolution which might be associated with supernovae.

I   Gravitational Collapse of Evolved Stars

FOR MORE than thirty years the concept of gravitational collapse of massive stars at the endpoint of thermonuclear evolution has intrigued theorists. This model has provided the first testing ground for evolutionary calculations of explosion dynamics of stars (Colgate and White 1966). Subsequent investigations have been made by Arnett (1966, 1967), Schwartz (1967), and Ivanova, Imshennik, and Nadezhin (1967). Each of these calculations began with a stable model and evolved it through an unstable phase by numerical solution of the equations of hydrodynamics. For all evolved stars too massive to become white dwarfs, Colgate and White (1966) found that neutrinos and antineutrinos escaping from the hot central regions would deposit sufficient energy in the overlying matter to explode these outer layers and produce supernovae. However, Schwarzschild (1958) has pointed out that the stellar death rate for those stars whose main-sequence mass is above the Chandrasekhar limit is much higher than the observed supernova rate. *Very* high

rates of quasistatic mass loss are required for agreement. Colgate and White
suggested that the dense core that remained after the explosion might always
have a sufficiently low mass to form a stable neutron star rather than approach
the Schwarzschild singularity of general relativity. Arnett (1966, 1967) and
Schwartz (1967) have obtained a similar result for small mass objects*
($M \sim 2$ $M_\odot$). Schwartz used Einsteinian rather than Newtonian gravitational
theory in his calculation; because of an improved estimate of thermal pressure at high density, both authors found that the core density was never high
enough to cause significant deviations from Newtonian gravitation theory.
For more massive objects ($M \gtrsim 8$ $M_\odot$), the situation is different. Using
somewhat different approaches to the problem of neutrino energy transfer,
Arnett (1967) and Ivanova *et al.* (1967) concluded that such massive objects
would not explode in the manner described by Colgate and White (1966). In
order to clarify the physical concepts involved, we present a simple analytic
argument as to why energy deposition by neutrinos is not effective in causing
mass ejection in the collapse of massive objects ($M \gtrsim 8$ $M_\odot$).

We will begin at the point at which an extremely dense core composed
primarily of neutrons has been formed, and material from the collapsing star
is falling down on it. Let us first get an estimate of the neutrino temperature.

For a spherical shell, the gravitational potential energy is given by

$$-d\Omega = \frac{GM_r}{r} dM_r. \tag{1}$$

We assume that the core will have a uniform density ϱ_c throughout. We
know that nuclear matter is highly incompressible, so this assumption is not
unreasonable. Thus

$$r = \left(\frac{3M_r}{4\pi\varrho_c}\right)^{1/3}. \tag{2}$$

In the extreme case of material free-falling onto the core, the kinetic
energy will be equal to the potential energy released; i.e.,

$$\tfrac{1}{2}u^2 dM_r = -d\Omega \tag{3}$$

where u is the velocity with which matter falls onto the core. Substituting
from equations (1) and (2), we obtain for the velocity

$$u = \left[2GM_r^{2/3}\left(\frac{4\pi\varrho_c}{3}\right)^{1/3}\right]^{1/2}. \tag{4}$$

---

* The mass referred to here is that of the highly evolved central region of the star, and
may be considerably less than the total stellar mass.

The flux of kinetic energy into the core is $F_{KE} = 1/2\varrho_{PS}u^3$, where ϱ_{PS} is the pre-shock density; if we then assume this energy is not stored in the core, it must be carried away by neutrinos. If we approximate the neutrino emission as blackbody radiation, then the energy flux is $F_\nu = \frac{7}{8}\sigma T^4$. Setting $F_{KE} = F_\nu$, we obtain the following expression for the neutrino temperature:

$$T = \left[\frac{4M_r\varrho_{PS}}{7\sigma} (2G)^{3/2} \left(\frac{4\pi\varrho_c}{3}\right)^{1/2}\right]^{1/4}. \tag{5}$$

This equation can be expressed in more convenient form if we put the mass in solar mass units and measure densities relative to that of nuclear matter. Equation (5) becomes

$$T = 1.82 \times 10^{12} \left(\frac{m}{\eta}\right)^{1/4} (\varrho^*)^{3/8}\ °K \tag{6}$$

where $m = M_r/M_\odot$, $\eta = \varrho_c/\varrho_{PS}$, and $\varrho^* = \varrho_c/3 \times 10^{14}$ gm/cm^3. In order to continue, we must have numerical values for mass and density. These cannot be determined analytically because of the complexity of the hydrodynamics; consequently, the relevant equations must be solved numerically by computer. When this is done, we find as a very conservative estimate $\varrho_{PS} \gtrsim 5 \times 10^{12}$ g/cm^3 and $M_r = 0.1$ M$_\odot$. Then at $\varrho^* = 1$, the temperature will be about $3.6 \times 10^{11}\,°K$. At this temperature, the average energy per nucleon will be about 45 MeV, which is above the Fermi level, but we assumed that the neutron gas was degenerate. Nevertheless, this estimate of the temperature of the core is reasonably good, because as more material falls on the core, the increase in density due to the stronger gravity is counteracted by the decrease in density resulting from thermal pressure. We then obtain a slightly higher nucleon temperature of $T_9 = 400$.

Antineutrinos will be produced mainly in two-nucleon interactions such as $n + n \to p + \bar{\nu}_e + e^- + n$. Bahcall and Wolf (1965) have estimated the cross section for this reaction in the zero-temperature limit; from their result we can get a rough estimate of the time scale for antineutrino production. This upper limit is of the order of 5×10^{-7} sec which is much less than the collapse time (on the order of milliseconds). Neutrinos will also be produced quickly by reactions like $p + e^- \to n + \nu_e$.

What happens to the neutrinos (antineutrinos) that are produced? To obtain a *lower limit* on the opacity, we will assume that they will interact only with degenerate electrons. Outside the core, the cross section for electron-neutrino scattering will be (Bahcall and Wolf 1965)

$$\sigma = 2 \times 10^{-44} \frac{\omega e_f}{3} \tag{7}$$

where ω is the neutrino energy and e_f the electron Fermi energy, both in units of electron rest masses, in the limit $\omega \gg e_f \gg 1$. These results agree reasonably well with the more detailed calculations done by Hansen (1966) using a more realistic electron energy distribution. Next we need an estimate for the electron number density. We consider the Fermi level at which electrons would be captured by the most refractory nucleus (which is ^{4}He). This most extreme case yields a minimum value of $n_e \sim 6 \times 10^{34}/\mathrm{cm}^3$ which corresponds to a ratio of electron number density to baryon number density of, at most, 1/50. We then compute the mean free path for electron-neutrino scattering and obtain a value of $\lambda = 3 \times 10^5$ cm. The radius R of a cold neutron star in static equilibrium with no envelope will be 2×10^6 cm. Because we are dealing with a dynamic situation in which there is thermal pressure, the region in which the electron density is at least 6×10^{34} cm^{-3} will be much larger than in the cold, static case. From the numerical solutions we have found that $R \geq 1.3 \times 10^7$ cm.

Neutrinos made in the core will not escape without scattering. We find that $R/\lambda = 43$ for e-$\bar{\nu}$ scattering and approximately 130 for e-ν scattering. The neutrinos have a much higher energy than the electron Fermi level, so that they will lose some fraction of their energy in the interaction. Using the estimate of Bahcall (1964) we find an average fractional energy loss of 1/4 to 1/2 per scatter. If we follow the path of the neutrino out of the star, we find that although the cross section decreases as the energy is degraded, it still scatters many times. The neutrinos become thermalized and leave with energies of the order of 1–2 MeV. The time scale for a neutrino to escape is of the order of 1–2 milliseconds, which is about the same as the hydrodynamic time scale.

The neutrinos will deposit nearly all their energy in regions having a density between 10^{11} and 3×10^{14} g/cm^3. The deposited energy will produce several important effects. As the electron gas is heated up, the cross section for neutrino scattering will increase. The number density of electrons will increase. The nucleon gas will be heated up so that significant thermal pressure will result. These effects all tend to hinder the escape of neutrinos (antineutrinos). Numerical calculations indicate that the thermal structure set up in this manner does *not* heat up the outlying layers enough to overcome gravitational binding, and the collapse continues. Because the core *cannot* radiate the energy supplied by gravitational infall, it heats up. Eventually the temperature will rise to the point where muons are formed. These will decay producing muon-neutrinos which will not interact at all and can therefore escape, removing energy. At this point, the central temperature ceases to rise. The mass in the core continues to increase, and the system approaches the Schwarzschild singularity of general relativity.

Less massive objects would be able to eject their outer layers because their lower opacity enhances the energy transfer by electron-type neutrinos and antineutrinos, and sets up a thermal structure in which the outer regions *are* unbound by gravity. *The quantitative value for the maximum mass which can explode is uncertain because of the many crude assumptions needed to make the problem tractable.* This point has been discussed in more detail elsewhere (Arnett 1968).

II Light Emission by Exploding Stars

In order to estimate the mass in the emitting volume of a supernova, we develop a simple physical model. Assume that the pulse of light centered around maximum is due to the diffusion of imprisoned radiation from an expanding mass of gas. The diffusion time is $\tau_{\text{diff}} \sim R^2/\lambda c$ where R is the radius of the cloud of gas, c is the velocity of light, and $\lambda = 1/\varkappa\varrho$ is the mean free path for the radiation. Near peak light this time scale should equal that of expansion, $\tau_{\text{exp}} \sim R/v$ where v is the velocity of expansion. This gives the relation $R\varrho = c/\varkappa v$ where $\varkappa$ is the opacity and ϱ the density. Assuming for simplicity that the density is more or less uniform through the cloud, the mass of the cloud is given by $M \sim 4\varrho R^3$. Eliminating ϱ, we find $M \sim 4R^2 c/\varkappa v$. Taking the Thomson scattering opacity, $\varkappa = 0.2 \text{ cm}^2/\text{gm}$, and an observed supernova expansion velocity, $v \sim (6 \text{ to } 10) \times 10^8$ cm/sec, only the radius remains to be determined. From the expansion velocity v and the time interval t from initial explosion to peak light, the radius may be calculated from $R \sim vt$. From the photographic light curves of Minkowski (1964), $t \sim 20$ to 50 days or about $(2 \text{ to } 5) \times 10^6$ sec, so that $R \sim (1.2 \text{ to } 5) \times 10^{15}$ cm.

Finally, we find $M/\text{M}_\odot \sim 0.7$ to 7.0, which seems to agree with the exploding star hypothesis. The smaller value $M \sim 0.7 \text{ M}_\odot$ suggests that the optical phenomena might be produced by a comparatively small mass spread over a large volume, and not by most of the material in the exploding star. It may be that the optical phenomena depends sensitively on pre-supernova

structure and the amount of gas in the vicinity. It is by no means obvious
that all exploding stars must necessarily behave in such a way in the visible
that they correspond to the optical phenomena called supernovae; nevertheless, Tremble's (1968) recent analysis of velocity vectors in the Crab nebula
strongly suggests that this supernova was an exploding star. The supernova
rate may only be a lower limit to the rate at which stars explode.

III   Evolutionary Models of the Final Stages of Stellar Evolution

In the previous two sections it has been suggested that theoretical models of
gravitational collapse and of supernova light output may be dependent on
the details of the structure of the star prior to becoming unstable. This section will describe some recent models of the oxygen and silicon-burning
stages of evolution of massive stars. Evolution of these models has been
followed to the point of hydrodynamic instability.

In order to reduce the exceedingly large amount of computer time needed
to correctly follow the evolution of a star possessing H, He, C, Ne, O, and

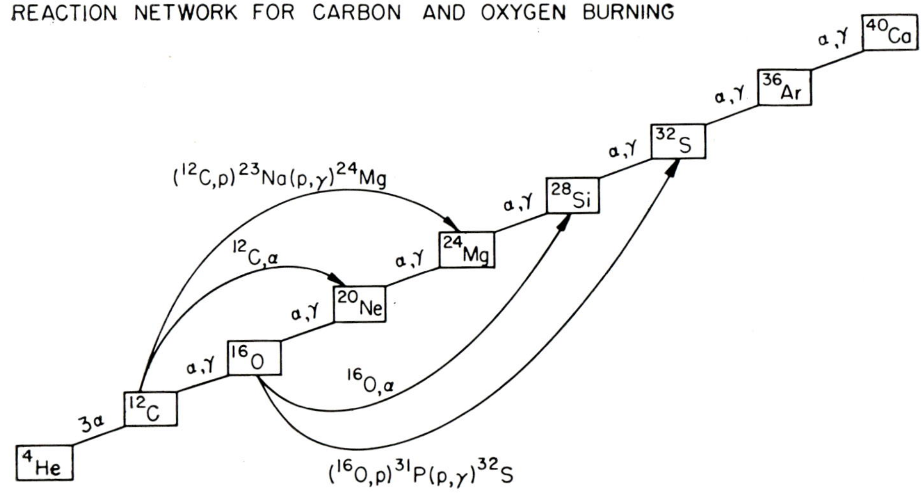

Fig. 1   Thermonuclear reaction network used for oxygen burning.

Si-burning shells, we consider the evolution of a pure ^{16}O star of 8 M$_\odot$. It is hoped that this model will correctly approximate the central regions of an evolved star. Evolutionary sequences are generated by numerical integration of the coupled equations of hydrodynamics, energy conservation and nuclear burning. The technique used is a combination of the two difference methods described by Sears and Brownlee (1965).

The change in composition is treated by a new numerical technique developed by Arnett and Truran (1968). At the extreme temperature encountered during oxygen and silicon burning, many reactions link a large number of nuclei. Because of this strong coupling between many nuclei an implicit scheme must be used. The reaction network chosen for oxygen burning is shown in Fig. 1. For silicon burning a network was chosen which consisted of the most important reaction links determined from the detailed analysis of Truran, Cameron, and Gilbert (1966). Reaction rates were taken from Truran, Hansen, Cameron, and Gilbert (1966) and Reeves (1965). The more recent summary of Fowler, Caughlan, and Zimmerman (1967) was not available when this evolutionary sequence was begun. Figure 2 shows the more complicated reaction network used for silicon burning.

Although simple analytic expressions are sufficient to represent the contributions to the equation of state from ions and radiation, the situation for electrons is more complex. The Fermi-Dirac integrals have been numerically evaluated to give a tabular equation of state which includes the effects of pair production, special relativity and quantum-statistical degeneracy. Interpolation of variable order allows the equation of state to be reproduced in an efficient manner with maximum error of less than 0.2 percent. Such high accuracy is desired because the evolutionary path leads to regions in which neutral dynamic stability is approached ($\gamma \sim 4/3$), and small errors might distort the structure significantly. Figure 3 shows deviations of the electron equation of state from an ideal gas ($P = \mathscr{R}\varrho T/\mu$), the regions of degeneracy and of significant formation of electron-positron pairs, and the boundary below which energy transfer by electron conduction is more effective than by radiative diffusion.

When temperatures above 0.5×10^9 °K are encountered, energy loss by neutrino emission is dominant. Figure 4 illustrates the extreme acceleration of the evolution of an 8 M$_\odot$ stellar core caused by the onset of substantial neutrino emission. The variation of radius of various mass fractions $q = M(r)/M$ with time is shown. The most rapid contraction during this stage occurs on a time scale which is of the same order as the pulsational period of

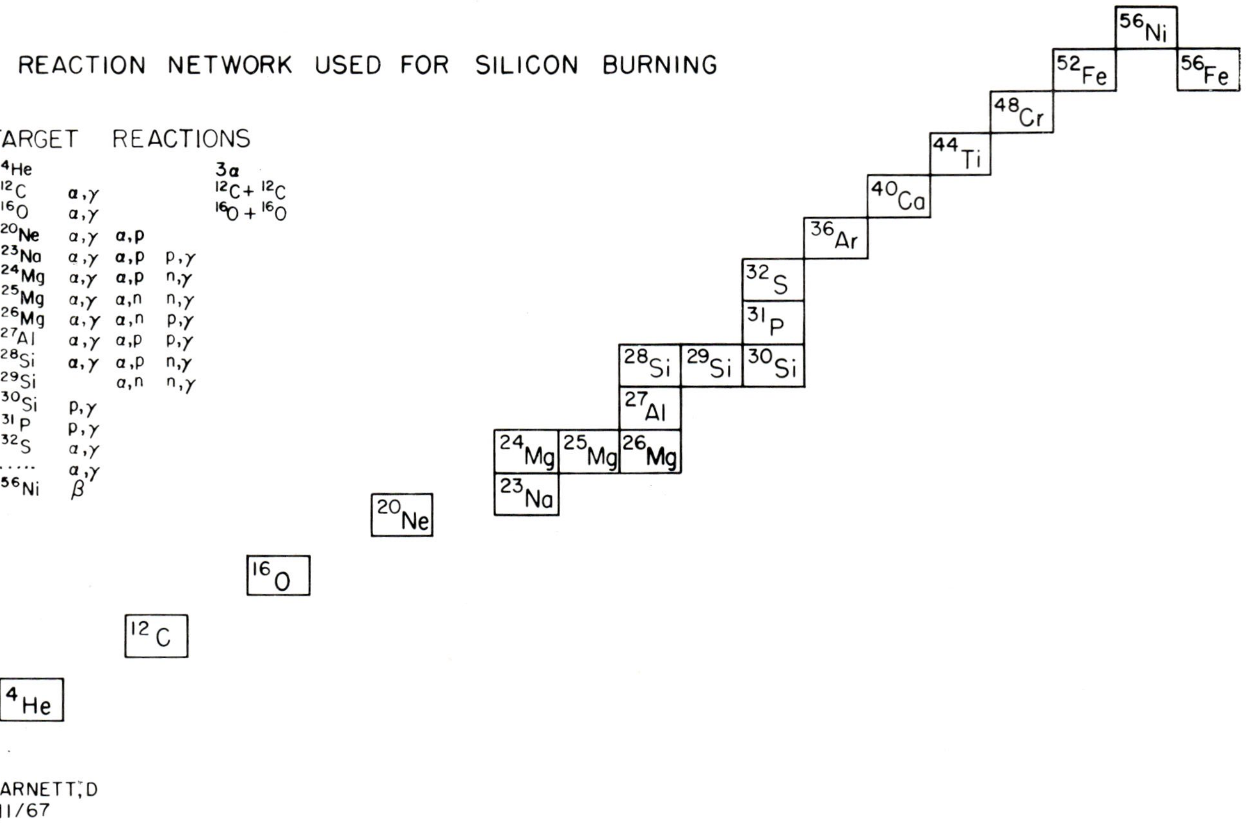

Fig. 2 Thermonuclear reaction network used for silicon burning.

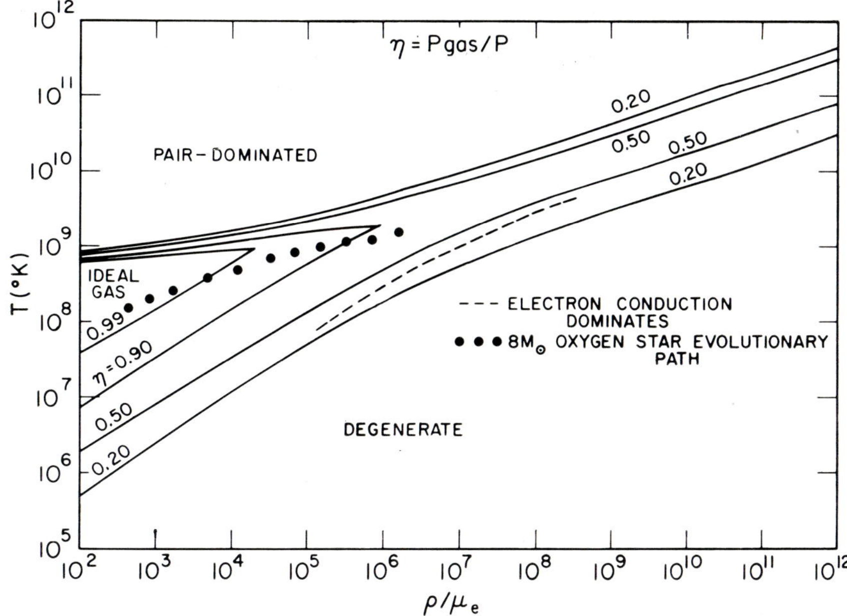

Fig. 3 Equation of state for electrons. The region of temperature and density for which thermally-formed electron-positron pairs dominate the equation of state is shown, as is the region in which quantum-statistical degeneracy is strong. The dominant mode of energy transport is by electron conduction below the dashed line. Deviations of the pressure from ideal gas behavior ($P_{gas} \equiv \mathscr{R}T\varrho/\mu_e$) are shown by contours of $\eta = P_{gas}$ (actual pressure).

a red giant envelope. In the core, contraction velocities are always much less than the local adiabatic sound speed so that the hydrostatic approximation is always valid during this stage.

Because electron degeneracy has not yet been encountered, the energy lost by neutrino emission must be supplied by gravitational contraction. This results in a somewhat more isothermal structure and a general increase in temperature. Eventually the $^{16}O + {}^{16}O$ reaction ignites in the center, releasing energy. Because of its higher dependence on temperature (Fowler and Hoyle 1964) the nuclear energy generation soon outstrips the neutrino energy loss locally. Under these conditions of high temperature and density both radiative diffusion and electron conduction are unable to transfer significant energy during the short time scale available. As was pointed out by Rakavy,

Shaviv, and Zinamon (1967) convection is the effective mode of energy transfer. The star adjusts to a static structure in which nuclear energy produced in a hot central region is transported outward by convection to a larger and cooler region to replace energy radiated by neutrino processes. The

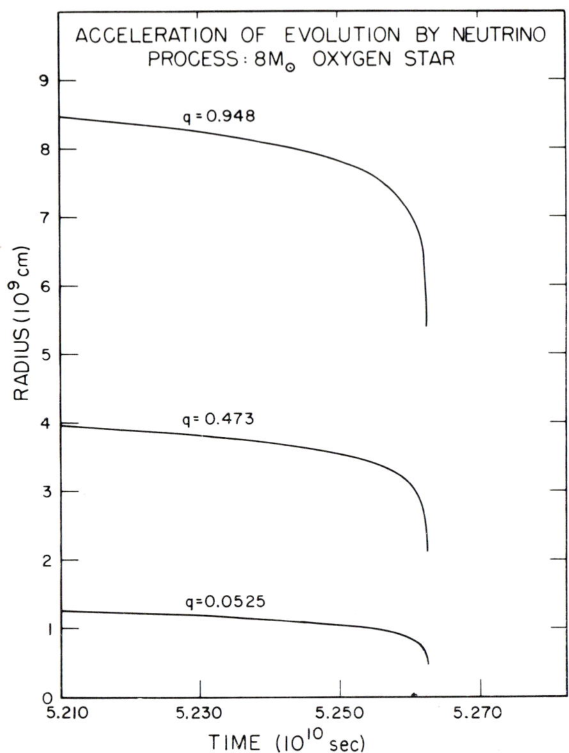

Fig. 4   Acceleration in the evolutionary change of structure caused by energy loss due to neutrino emission.

mixing-length theory of convection (Spiegel 1963) was used. Figure 5 displays the rate of energy loss by neutrino emission and of energy gain by nuclear reactions as a function of mass fraction q for the stage of oxygen burning in the center of the star. The excess of nuclear energy generation in the center of the star and its more rapid decrease moving away from the center are clearly shown. The "humps" in both the rate of nuclear energy generation and of neutrino energy loss are due to a less dramatic "hump" in the temperature distribution.

The central temperature and density are shown in Fig. 6 from the initial stage of contraction induced by neutrino emission to the final stage calculated, also a contraction, in which endoergic nuclear photodisintegration occurs in the center. Several interesting stages of the evolution have been selected and

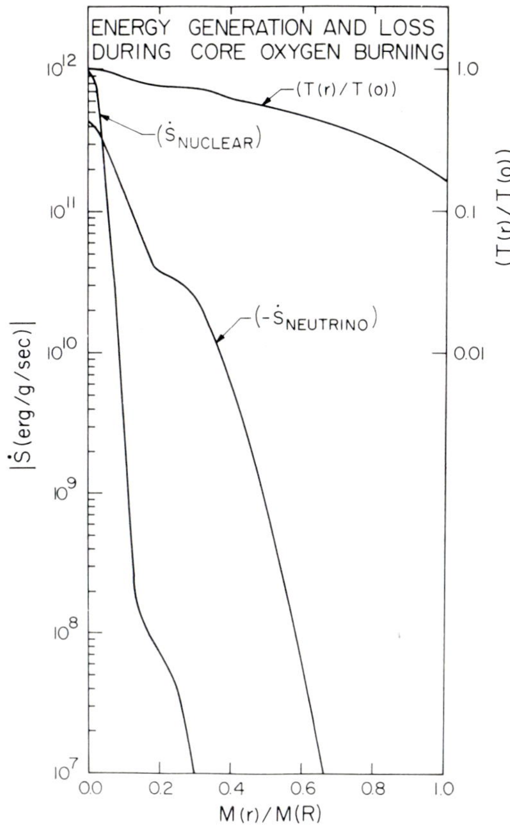

Fig. 5   Energy generation by $^{16}O + {}^{16}O$ reactions and energy loss by neutrino emission as a function of position in the star.

labeled with the letters A through J. Also shown is the curve (T^3/ϱ = constant) which the center would follow in the absence of neutrino and nuclear effects. Notice that the center tends to move toward regions of lower entropy (smaller T^3/ϱ).

  W. D. ARNETT

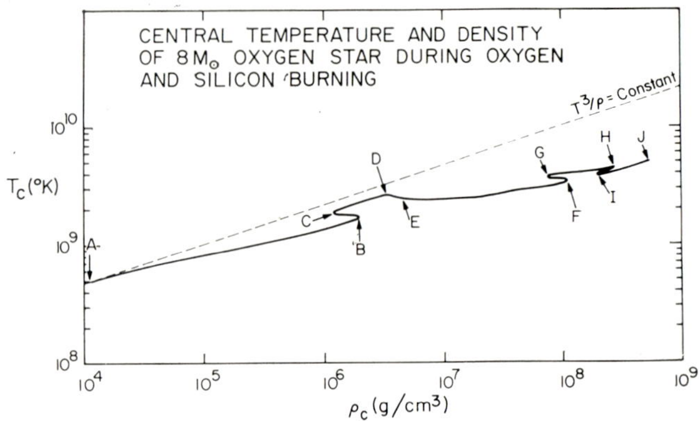

Fig. 6 Temperature and density at the stellar center during the stages of oxygen and silicon burning.

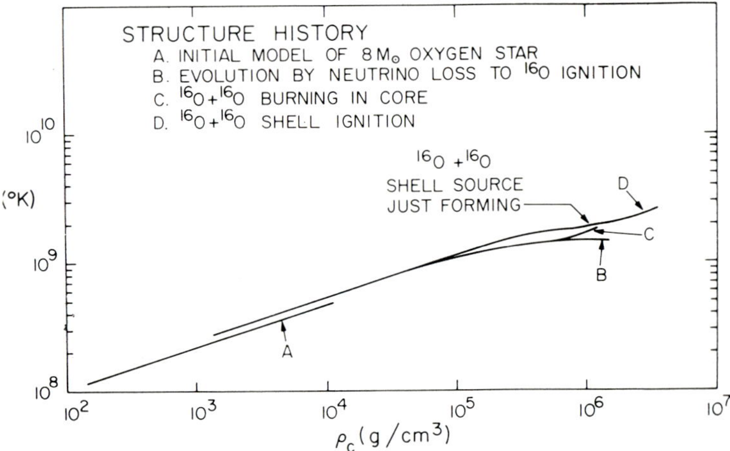

Fig. 7 "Snapshots" of temperature and density structure at selected instants in time: From initial model through ^{16}O + ^{16}O shell ignition.

The run of temperature and density throughout the entire star at several selected instants of time ("snapshots") are shown in Figs. 7 through 9. The identifying letters correspond to those in Fig. 6 and Table 1. We will now discuss each of these models briefly.

Table I Physical parameters of selected models

Model	$T_c/10^9\,°K$	ϱ_c	$-\varepsilon_\nu$	ε_{nuc}	Γ	Kinetic energy	Binding energy	Time difference	Comments
		(g/cm^3)	(erg/g-sec)	(erg/g-sec)		(erg)	(erg)	(sec)	
A	0.484	1.14 (+4)	1.52 (+5)	—	—	1.45 (+31)	3.37 (+50)		Initial model
B	1.680	1.97 (+6)	6.49 (+10)	1.103 (+11)	—	3.14 (+39)	9.80 (+50)	5.26 (+10)	^{16}O core ignition
C	1.884	1.20 (+6)	3.98 (+11)	7.17 (+11)	—	3.93 (+39)	1.01 (+51)	1.03 (+6)	^{16}O core burning
D	2.616	3.52 (+6)	5.38 (+12)	2.94 (+12)	—	3.83 (+39)	1.25 (+51)	2.05 (+5)	^{16}O shell ignition
E	2.426	5.17 (+6)	1.58 (+12)	3.42 (+10)	—	2.45 (+40)	1.20 (+51)	7.50 (+3)	Core cooling after ^{16}O burning
								2.21 (+5)	
F	3.476	1.10 (+6)	9.93 (+11)	3.06 (+12)	1.3919	6.21 (+40)	1.21 (+51)		Si core ignition
G	3.594	7.70 (+7)	3.17 (+12)	4.56 (+12)	1.3835	7.45 (+41)	1.17 (+51)	2.95 (+4)	Si core burning
H	4.503	2.79 (+8)	3.24 (+12)	6.57 (+10)	1.3890	1.05 (+42)	1.22 (+51)	5.45 (+4)	Si shell ignition
I	3.900	2.02 (+8)	1.06 (+12)	5.72 (+10)	1.3798	5.66 (+41)	1.17 (+51)	1.24 (+4)	Core contraction resumes
								7.22 (+3)	
J	5.078	5.38 (+8)	3.21 (+12)	−3.45 (+15)	1.3716	2.41 (+48)	1.23 (+51)		Photodisintegration of iron peak

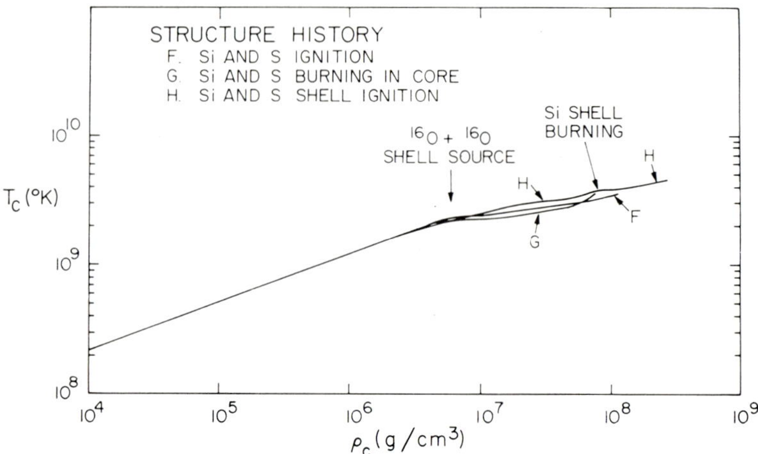

Fig. 8 "Snapshots" of temperature and density structure at selected instants in time: From ignition of silicon burning in the core through shell ignition of silicon burning.

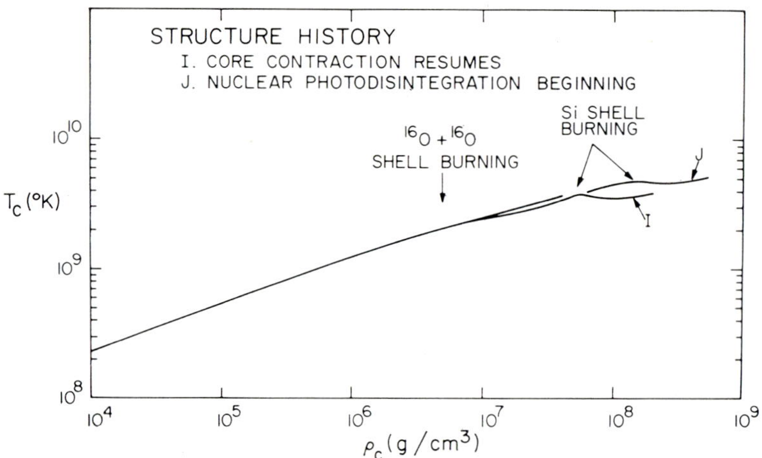

Fig. 9 "Snapshots" of temperature and density structure at selected instants in time: Final stages to initiation of dynamic collapse.

A) *Initial model of an 8 $M_\odot$ oxygen star*   The initial composition was pure ^{16}O and the initial pressure-density structure was that of a polytrope of index 3. At He-depletion in the center of the 30 $M_\odot$ stellar model of Stothers (1966) the products of helium burning comprised the inner 11.1 $M_\odot$

of the star; model A can be envisaged as a crude extrapolation of Stother's evolutionary sequence. Initially in radiative equilibrium at a sufficiently low central temperature ($\sim 0.5 \times 10^9\,°K$) that there is no nuclear burning, the star is forced to contract at an ever increasing rate as was shown in Fig. 4.

B) *Evolution by neutrino emission to oxygen ignition*   During this contraction an interesting change in structure occurs. The neutrino emission in the center is stronger than energy generation by gravitational contraction so that the entropy of that region decreases. In the middle and outer regions, the

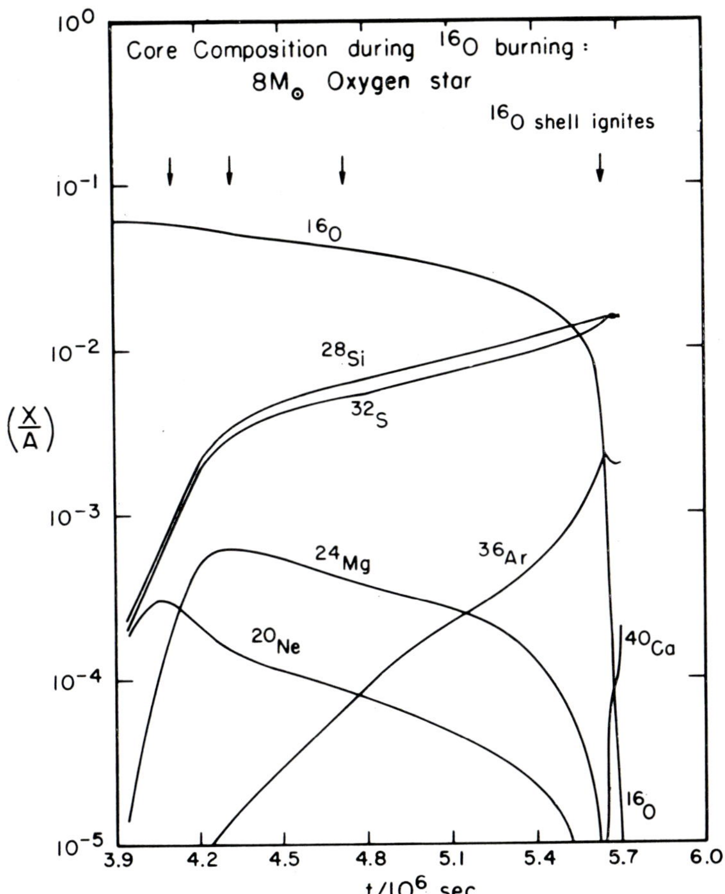

Fig. 10   Change in nuclear composition of the core during oxygen burning.

opposite occurs. The gravitational energy generation is greater than the neutrino energy loss so that the entropy increases. This effect may be seen by comparison of models A and B in Fig. 7. The evolution from A to B required 5.26×10^{10} seconds or 1660 years. This is by far the most lengthy stage of evolution encountered. As may be seen from Table 1, subsequent evolution takes less than 18 days.

C) $^{16}O + {}^{16}O$ *burning at the stellar center* Eventually the increase in temperature is sufficient to ignite the $^{16}O + {}^{16}O$ reaction at the center. Because this nuclear reaction rate increases more rapidly with temperature than does that of neutrino emission, the central temperature begins to rise. The temperature gradient near the center continues to steepen until it can drive convection. The central convective region grows until it reaches $1.6\,M_\odot$. As may be seen in Fig. 7, the extent of the convection zone in model C is determined by the thermal structure. The nearly isothermal region due to neutrino emission has a significant entropy gradient which is positive. This inhibits the growth of the convective core, restricting it to only 20 percent of the mass.

The composition of the core during oxygen burning is shown in Fig. 10. The main products of oxygen burning are ^{28}Si and ^{32}S with some ^{36}Ar. The variation of temperature and density during oxygen burning are significant, as is illustrated in Fig. 11. Temperature variations are particularly large during final stages of oxygen depletion.

D) $^{16}O + {}^{16}O$ *shell burning* Eventually depletion of ^{16}O in the convective core results in termination of oxygen burning in the center, and an ^{16}O source forms. Neutrino cooling reduces the central temperature gradient below adiabatic and the central density increase.

E) *Core cooling after oxygen burning* As convection dies out in the center, the central temperature begins to decrease. This is followed by an almost isothermal increase in central density to $\varrho \sim 1.1 \times 10^8$ g/cm³. The oxygen shell source develops a convective region which grows until it contains $5.74\,M_\odot$. Because of the higher entropy at which the shell burning occurs, it can overcome the positive entropy gradient which inhibited the growth of the core convective zone.

F) *Ignition of Si in the core* As the core contracts to high densities the gravitational energy generation eventually surpasses the cooling by neutrino emission and the central temperature rises high enough to begin the complex

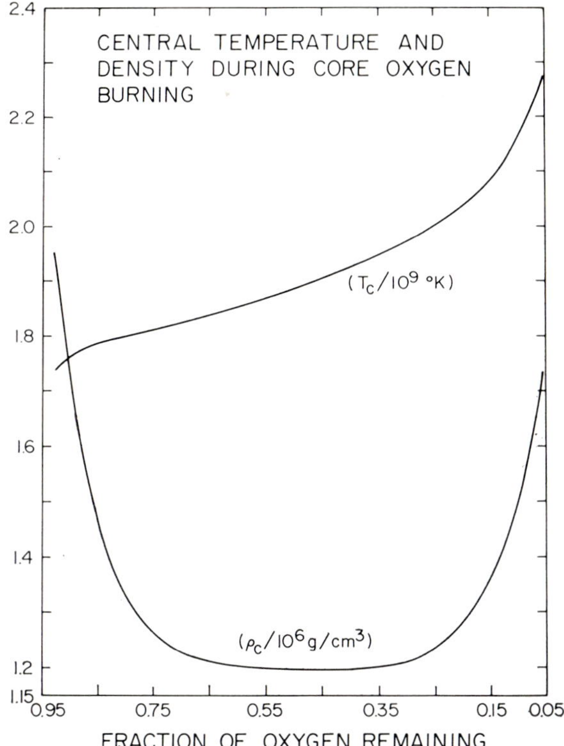

Fig. 11 Variation of central density and temperature during oxygen
burning in the core.

network of photodisintegration and capture reactions called silicon burning.
The central temperature increases and central density decreases.

G) *Silicon burning in the core*   As silicon burning becomes more effective
in the core it develops a small convective zone in the center. As in core
oxygen burning, the extent of this convective zone is limited by a surrounding isothermal region. The oxygen-burning shell is very active, extending its
convective region almost to the "surface" of the model. Figure 12 shows the
composition change in the core during silicon burning. The more complex
nature of this process is apparent. In this calculation the positron decay of
^{56}Ni was ignored although the time scale was sufficiently long to allow some
decay to occur. Figure 13 shows the pronounced variation of central temperature and density during core silicon burning.

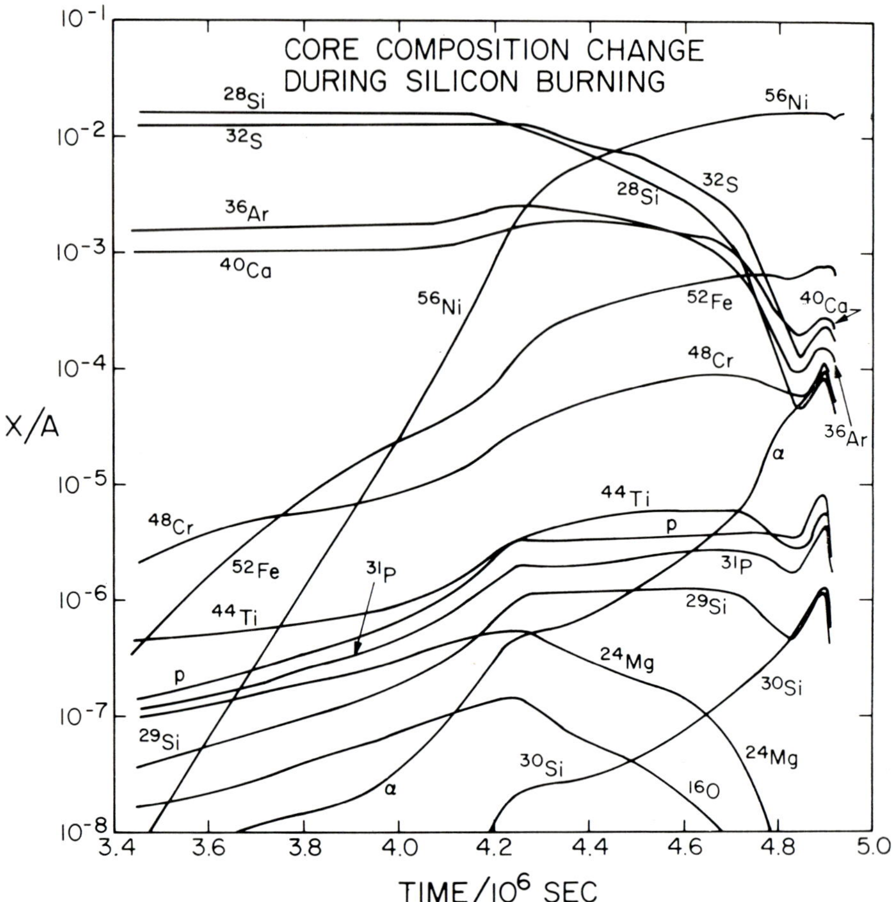

Fig. 12   Change in nuclear composition of the core during silicon burning.

H) *Ignition of silicon shell burning*   When the convective core is converted to ^{56}Ni, a silicon-burning shell is formed in a manner similar to that described for oxygen burning.

I) *Core contraction*   After the finish of silicon burning in the core a core contraction begins. Because this region is approaching neutral stability ($\Gamma \gtrsim 4/3$) the contraction easily goes to higher density with negligible effective energy production by gravitational contraction.

J) *Nuclear photodisintegration*   The central temperature rises until the nuclear equilibrium favors an appreciable number of free protons and alpha

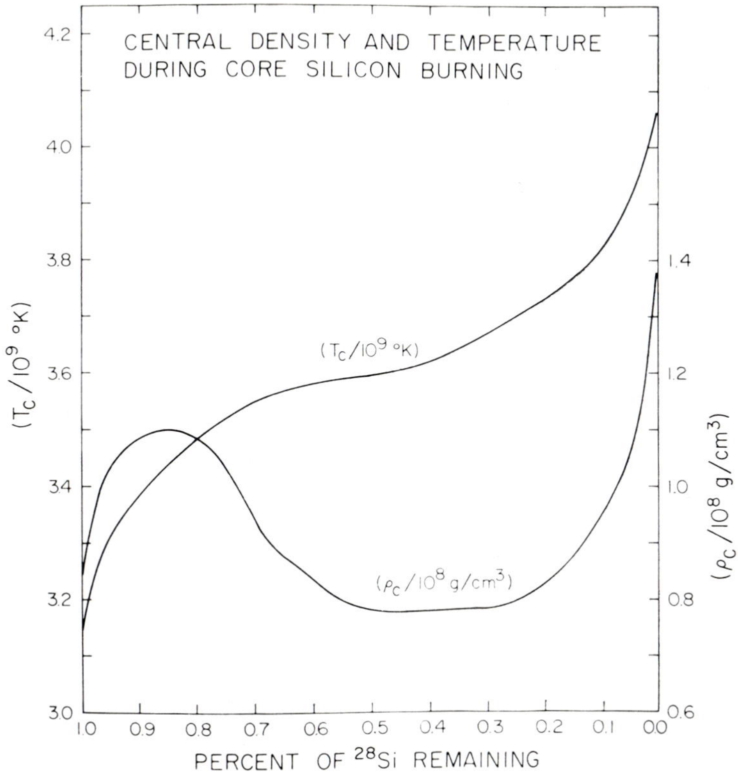

Fig. 13 Variation of central density and temperature during silicon burning in the core.

particles. The reactions which effect this change are endoergic. The resulting energy sink is quite large (a thousand times the core neutrino loss rate of model J) and drastically speeds the core contraction. Evolution now occurs on a hydrodynamic time scale. In 7.22×10^3 sec the kinetic energy increases from $\sim 10^{-10}$ to $\sim 2 \times 10^{-3}$ of the gravitational binding energy of the model.

IV Comments

There are several ways in which the evolutionary sequence just described could be improved. (1) During the later stages the characteristic time scale for evolutionary change was of the same order as the time required for a convective element to transverse a pressure scale-height. This suggests that a dynamic theory of convection be used, especially during the later stages. (2) Mass zoning was crude, as befits an exploratory calculation. After silicon burning in the core this zoning was too crude to be trusted in detail; these last stages should be recalculated. (3) Neglect of H, He, C, and Ne-burning stages may not be justified. (4) The effect of electron captures, especially $e^- + p \to \nu + n$, on instability and nucleosynthesis after silicon burning should be examined. (5) The rate of the $^{16}O + ^{16}O$ reaction is very uncertain; an experimentally determined cross section is needed. Nevertheless the evolutionary sequence just described does attempt to examine the evolution of the late stages of stars with a more realistic treatment of the nuclear burning stages than has been possible before. Such model sequences enable us to examine the initial stages of gravitational collapse in a quantitative manner and to connect conventional evolutionary sequences to dynamic instabilities. They may allow us to ascertain whether the thermonuclear explosion (Fowler and Hoyle 1964) or the neutrino transfer (Colgate and White 1966; Arnett 1966, 1967) mechanism is relevant to supernovae. Of particular interest is the formation of a small (1.6 $M_\odot$), unstable central region surrounded by a large mass of potentially explosive fuel. In a previous investigation of the collapse of stellar cores (Arnett 1967), it was concluded that for "stellar cores" composed of iron-peak elements there existed a maximum mass $4\,M_\odot \lesssim M \lesssim 8\,M_\odot$ which could expel overlying material by neutrino (antineutrino) energy transport. However, in §I we saw that this upper limit to the mass should be lowered because we used a *lower* limit for the neutrino (antineutrino) opacity. It is therefore uncertain whether or not neutrino (antineutrino) energy transport will be effective for exploding the 1.6 $M_\odot$ "stellar core" of model J. This question may be clarified by analysis of the onset and development of hydrodynamic instability, including some of the improvements listed above.

References

Arnett, W.D. 1966, *Can. J. Phys.*, **44**, 2553
– 1967, *Can. J. Phys.*, **45**, 1621.
– 1968, *Ap. J.* **153**, July, in press.
Arnett, W.D. and Truran, J.W. 1968, "Carbon-Burning Nucleosynthesis at Constant Temperature" (preprint).
Bahcall, J.N. 1964, *Phys. Rev.*, **136**, B1164.
Bahcall, J.N. and Wolf, R.A. 1965, *Phys. Rev.*, **140**, B1452.
Colgate, S.A. and White, R.H. 1966, *Ap. J.*, **143**, 626.

Fowler, W. A., Caughlan, G. R., and Zimmerman, B. A. 1967, *Ann. Rev. Astr. and Astrophys.*, **5**, 525.

Fowler, W. A. and Hoyle, F. 1964, *Ap. J. Suppl. 91*, **9**, 201.

Hansen, C. J. 1966, Ph.D. dissertation, Yale University, unpublished.

Ivanova, L. N., Imshennik, V. S., and Nadezhiv, D. K. 1967, "Investigation of the Dynamics of the Explosion of Supernovae" (preprint).

Minkowski, R. 1964, *Ann. Rev. Astr. and Astrophys.*, **2**, 247.

Rakavy, G., Shaviv, G., and Zinamon, Z. 1967, *Ap. J.*, **150**, 131.

Reeves, H. 1965, *Stellar Structure*, ed., L. H. Aller and D. B. McLaughlin (Chicago: University of Chicago Press) p. 113.

Schwartz, R. A. 1967, *Ann. Phys.*, **43**, 42.

Schwarzschild, M. 1958, *Structure and Evolution of the Stars* (Princeton: Princeton University Press).

Sears, R. L. and Brownlee, R. R., 1965, *Stellar Structure*, ed. L. H. Aller and D. B. McLaughlin (Chicago: University of Chicago Press) p. 575.

Spiegel, E. A., 1963, *Ap. J.*, **138**, 216.

Stothers, R., 1966, *Ap. J.*, **143**, 91.

Tremble, V., 1968, Ph.D. dissertation, California Institute of Technology, unpublished.

Truran, J. W., Cameron, A. G. W., and Gilbert, A., 1966, *Can. J. Phys.*, **44**, 563.

Truran, J. W., Hansen, C. J., Cameron, A. G. W., and Gilbert, A., 1966, *Can. J. Phys.*, **44**, 151.

3) *Supernova explosions*   The difficulties encountered in the attempts to account for the observed heavy element abundances in the galaxy by either of the previous mechanisms tend to support the view that the main source of heavy element synthesis is stellar evolution through the entire history of the galaxy, with gradual enrichment of the interstellar gas by the material ejected in supernova explosions. Further, the generally endoenergetic character of the various processes of heavy element production suggest that the supernova event might itself be the site of substantial heavy element production.

Recent studies of the dynamics of the cores of highly evolved massive stars (Colgate and White, 1966; Arnett, 1966, 1967a) have revealed that for cores of mass $M_c \lesssim 4\ M_\odot$ a considerable fraction of the core mass will be ejected. This mass may be identified with supernova explosions. At the extreme temperature and density conditions realized in the core the matter will be composed predominantly of neutrons. As the ejected matter expands and cools, these neutrons may be converted to protons, neutrons and heavier elements. Furthermore, the high temperatures and densities imparted to the surrounding envelope by the passage of the resulting shock wave can result in heavy element synthesis in these regions.

This paper will be concerned with the implications of these supernova models for the synthesis of the heavy elements. The outline of this paper is as follows. In Section II, the general problem of the approach to nuclear statistical equilibrium will be discussed as it relates both to stellar evolutionary models and to presupernova conditions. This will be followed in Section III by a discussion of the various nuclear transmutation processes which might take place in the supernova envelope in the wake of the shock wave with emphasis on the possible synthesis of iron peak elements in the regions immediately surrounding the supernova core. In Section IV, the implications of the extreme conditions realized in the supernova core for the synthesis of heavier elements by rapid neutron capture will be considered.

II   The Equilibrium Process

The notion that a thermodynamic equilibrium process might explain many aspects of the observed abundance features dominated early work on the subject of nucleosynthesis (see the review article by Alpher and Herman, 1950). While these early equilibrium theories encountered many difficulties and were ultimately discarded, current theories attribute the main features of the iron abundance peak to such a process (Hoyle, 1946). Burbidge *et al.* (1957) adopted this point of view and proceeded to determine (by means of a consideration of various elemental and isotropic abundance ratios) the physical conditions which provided the best fit to the observed abundances of the iron group nuclei. A typical equilibrium fit to the observed iron peak abundances is shown in Fig. 1. While this work largely substantiated the role

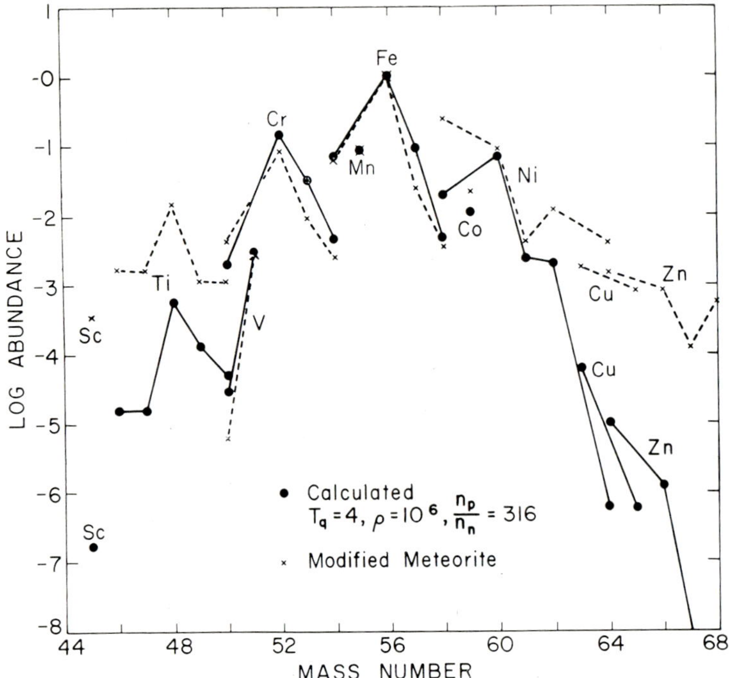

Fig. 1   The abundance relative to ^{56}Fe of nuclei formed by the equilibrium process (after Cameron, 1963). The high meteoritic iron abundance has been reduced by a factor 5.

of an equilibrium proces in the synthesis of these elements, the details of the thermonuclear reaction sequence which leads to the equilibrium condition and thus of the lifetime of this phase of stellar evolution were not available.

The subject of silicon burning and the ensuing approach to nuclear statistical equilibrium has recently been treated in detail by several authors (Truran, Cameron and Gilbert, 1966; Bodansky, Clayton and Fowler, 1968a and b; Truran, 1968a and b). In the calculations performed by Truran *et al.*, the non-linear system of equations for the time rates of change of the abundances of the various nuclear species were solved by means of a backward differencing procedure (Arnett and Truran, 1968). In contrast, Bodansky *et al.* determined the abundances of the heavier elements assuming them to be instantaneously in equilibrium with respect to ^{28}Si and the free densities of light particles—neutrons, protons and α-particles. This assumption may be justified qualitatively by the fact that the effective photodisintegration rate of ^{28}Si, governed in large measure by the photodisintegration rate of ^{24}Mg, is slow compared to the typical reaction times for neutron and charged-particle reactions proceeding on the heavier nuclei. The range of validity of this approximation has been considered in a recent paper by this author (Truran, 1968a). It should be pointed out that for those conditions for which the "quasi-equilibrium" approximation is valid, this method provides an extremely rapid procedure for determining the iron-peak abundances.

In the following paragraphs, the results of "typical" calculations of the approach to equilibrium will be presented. It should be emphasized that these results are relevant not only to the problem of the formation of iron-peak elements in supernova shock waves (see Section III) but also to considerations of the final presupernova phase of evolution in more massive stars.

The nuclei included in this study are shown in Fig. 2. Including neutrons, protons and α-particles, a total of 82 nuclear species were considered. All (n, γ), (p, n), (p, γ), (α, p), (α, n), (α, γ) reactions and their corresponding inverses were incorporated into the rate equations as well as the triple-alpha reaction and the heavy ion reactions ^{12}C + ^{12}C and ^{16}O + ^{16}O. Where experimental data are available, the rates for these reactions are determined as sums over individual resonance contributions (Reeves, 1965; Fowler, Caughlan and Zimmerman, 1967; Truran *et al.*, 1966b); otherwise, they are calculated theoretically from average nuclear properties (Truran, Hansen, Cameron and Gilbert, 1966). Nuclear β-decay processes (electron decay, positron decay and electron capture) have also been included; the rates for these reactions are those calculated by Hansen (1966). The behavior of this nuclear reaction network is governed by the equations for the time rates of change of the nuclear abundances.

The evolution of the abundances in time for a region composed initially of pure ^{28}Si at a temperature of $3 \times 10^9\,^\circ$K and a density of 2×10^6 g/cc is shown in Fig. 3. The early stages of the evolution displayed here are char-

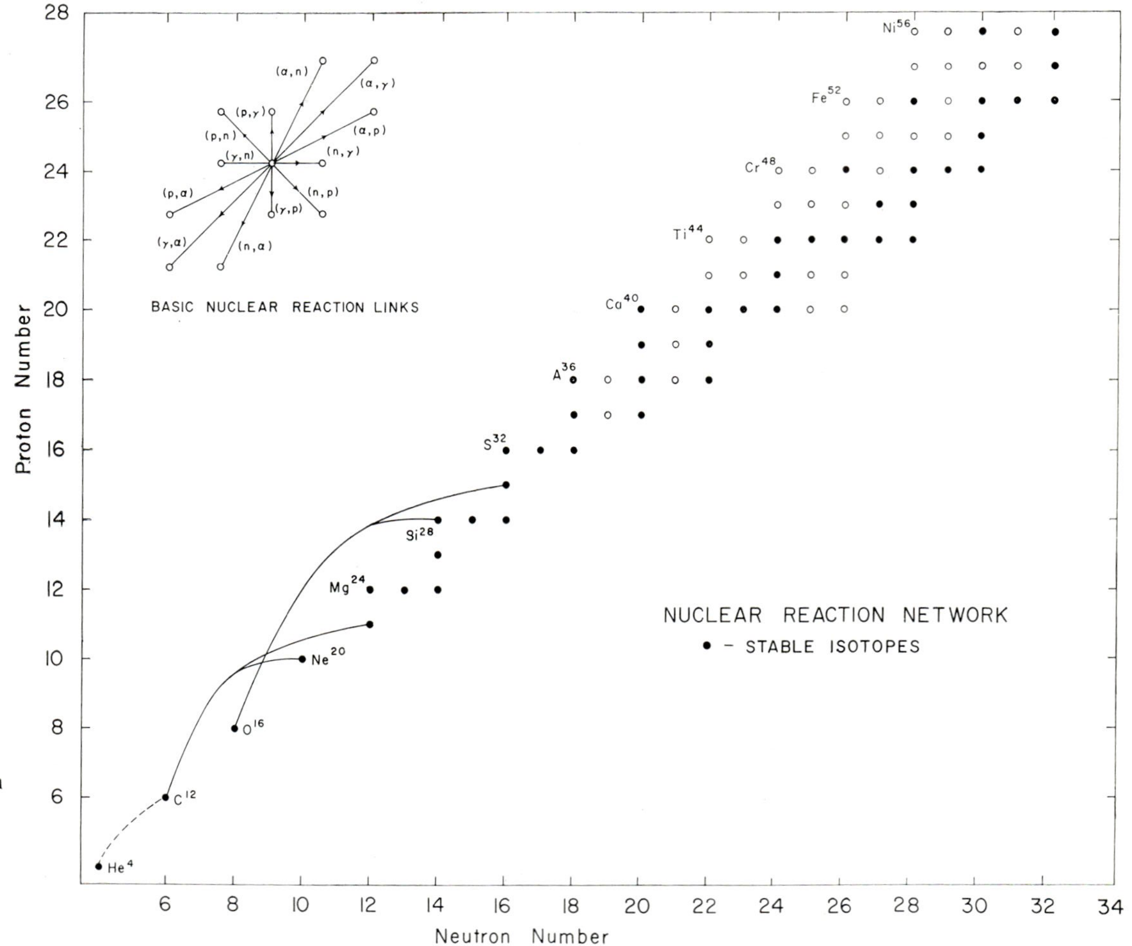

Fig. 2 Schematic of the nuclear reaction network employed for studies of silicon burning.

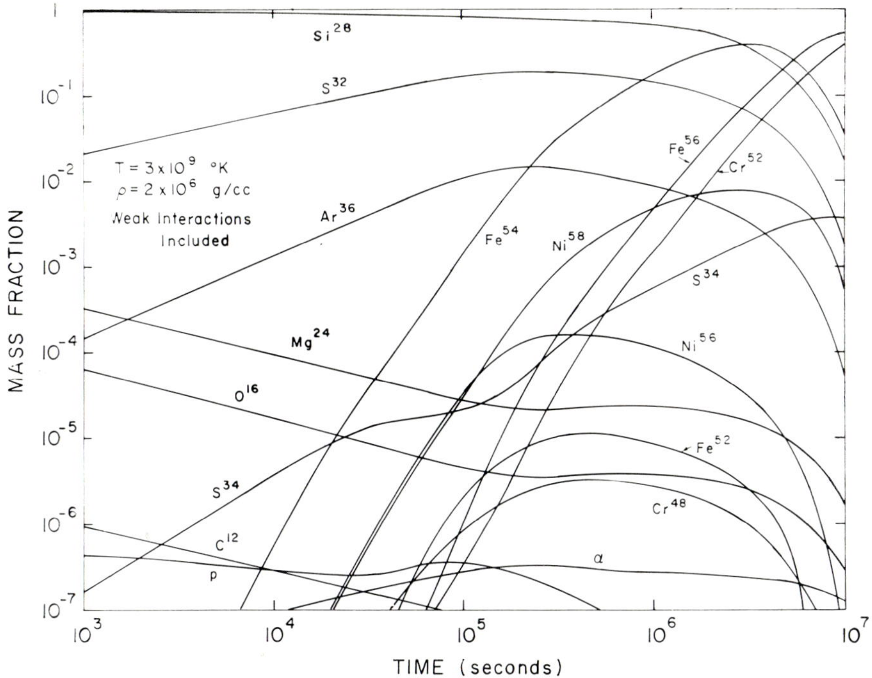

Fig. 3   The mass fractions of the more abundant constituents plotted as a function of time for silicon burning at $T = 3 \times 10^9$ °K, $\varrho = 2 \times 10^6$ g/cc.

acterized by the buildup of elements heavier than silicon toward an instantaneous equilibrium ("quasi-equilibrium") relative to ^{28}Si and alpha particles, protons and neutrons. On a time scale of 10^5 seconds, the dominant heavy nuclei formed are alpha-particle nuclei or alpha-particle nuclei plus two neutrons (i.e., ^{52}Fe, ^{54}Fe, ^{56}Ni and ^{58}Ni). When this silicon burning stage has gone to completion, however, the dominant nuclei are found to be ^{52}Cr and ^{56}Fe. This change in character of the abundance distribution results from neutron enrichment of the system by means of positron decay and electron capture processes.

The sensitivity of the abundance features to the degree of neutron enrichment is demonstrated by a comparison of these results with those of a similar calculation in which all nuclear beta-decay processes were omitted, as shown in Fig. 4. For this case, the most abundant product of silicon burning is found to be ^{56}Ni. The neutron enrichment responsible for the very different abundances realized in Fig. 3 was only $\sim 1\%$.

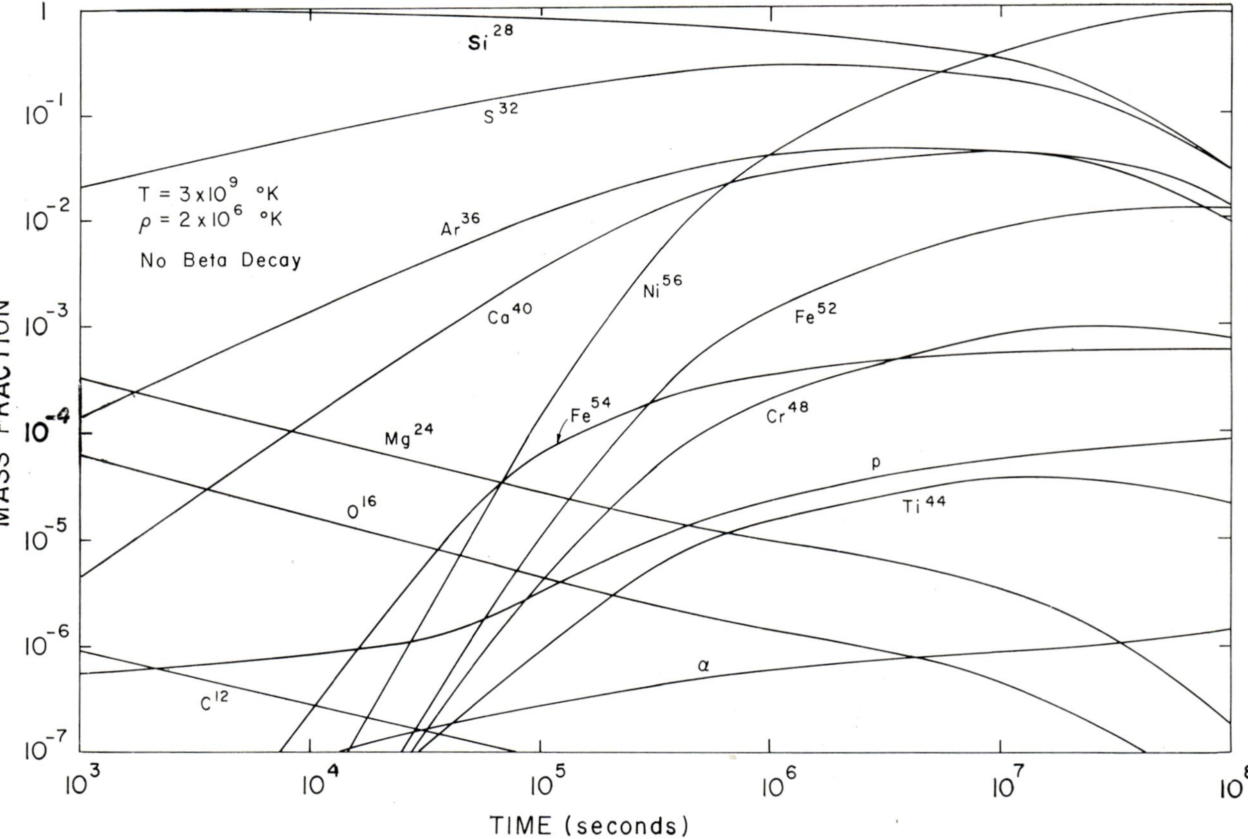

Fig. 4 Mass fractions plotted as a function of time for silicon burning at $T = 3 \times 10^9$ °K with weak interactions artificially turned off.

The evolution of the abundances in time for a region composed initially
of pure ^{28}Si for the conditions $T = 4 \times 10^9\,^\circ$K and $\varrho = 2 \times 10^7$ g/cc is shown
in Fig. 5. At this higher temperature, the lifetime of the silicon burning pro-
cess is reduced by a factor 10^4–10^5 and the influence of nuclear beta decay
processes is considerably lessened. The most abundant product after 100 seconds is found to be ^{56}Ni.

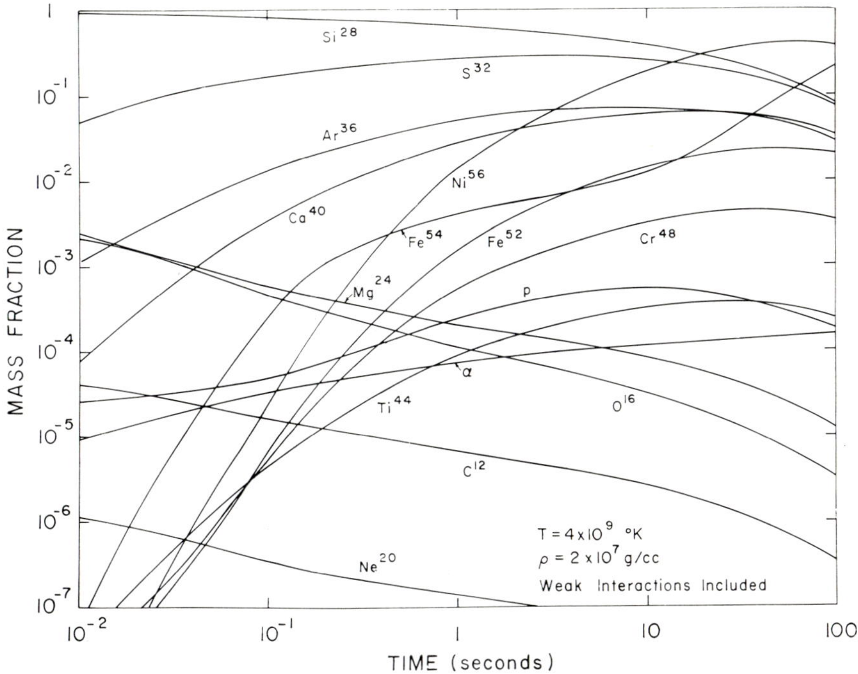

Fig. 5 Mass fractions plotted as a function of time for silicon burning
at $T = 4 \times 10^9\,^\circ$K, $\varrho = 2 \times 10^7$ g/cc.

The choice of a pure ^{28}Si composition at the onset of silicon burning is
clearly an approximation. Generally, the medium will consist of some distribution of the products of the proceeding carbon- and oxygen-burning phases,
the most abundant products being the alpha-particle nuclei ^{24}Mg, ^{28}Si and
^{32}S (Cameron, 1959 a and b; Reeves and Salpeter, 1959; Tsuda, 1963; Arnett
and Truran, 1968). ^{28}Si, having the highest separation energies for protons
and alpha particles, will be favored as the temperature is further increased,
although the abundances of ^{24}Mg and ^{32}S may still be significant.

In order to investigate the importance of admixtures of other nuclei in the medium at the onset of silicon burning, calculations were carried out for the conditions $T = 3 \times 10^9\,°K$, $\varrho = 2 \times 10^6$ g/cc for two further compositions:

1) the medium is composed of equal amounts by mass of ^{28}Si and ^{24}Mg

$$\chi(^{28}\text{Si}) = \chi(^{24}\text{Mg}) = 0.5$$

2) the medium is composed of equal amounts by mass of ^{28}Si and ^{32}S

$$\chi(^{28}\text{Si}) = \chi(^{32}\text{S}) = 0.5$$

Nuclear beta-decay processes have been purposely omitted from these calculations. The inclusion of these reactions under our conditions would have the effect of increasing the number of neutrons relative to protons in the system and therefore of altering the resulting equilibrium abundance distribution for a specified temperature and density. For a determination of the extent to which the silicon-burning process is accelerated by the presence of other intermediate mass nuclei, it is therefore convenient to ignore beta decay and to observe the approach to an identical nuclear statistical equilibrium abundance distribution.

The evolution in time of a region composed of equal concentrations by mass of ^{28}Si and ^{24}Mg at the onset of silicon burning is compared to the evolution of a pure ^{28}Si region in Fig. 6. These results clearly demonstrate that on a time scale of less than 10^3 seconds any appreciable abundance of ^{24}Mg will be transmitted into ^{28}Si, and the subsequent evolution of this silicon region will be indistinguishable from that for a region composed initially of pure ^{28}Si (see Fig. 4). This behavior results from the fact that the rate of alpha-particle capture on ^{24}Mg considerably exceeds that for the corresponding capture on ^{28}Si. Alpha particles released in the early stages of silicon burning are then rather completely absorbed in the conversion of ^{24}Mg to ^{28}Si until a condition of "quasi-equilibrium" is again approached. In this calculation, the evolution of a region composed initially of ^{28}Si and ^{24}Mg was found to become indistinguishable from the pure ^{28}Si case at time $t = 4 \times 10^2$ seconds, at which point the silicon mass fraction was $\chi(^{28}\text{Si}) = 0.987$.

The influence of a substantial concentration of ^{32}S in the medium at the onset of silicon burning is demonstrated in Fig. 7. Again, this evolution has been compared to that for a region composed initially of pure ^{28}Si. For this case, a time scale of the order of 10^7 seconds is required before the abundance pattern becomes indistinguishable from that for a pure ^{28}Si initial condition

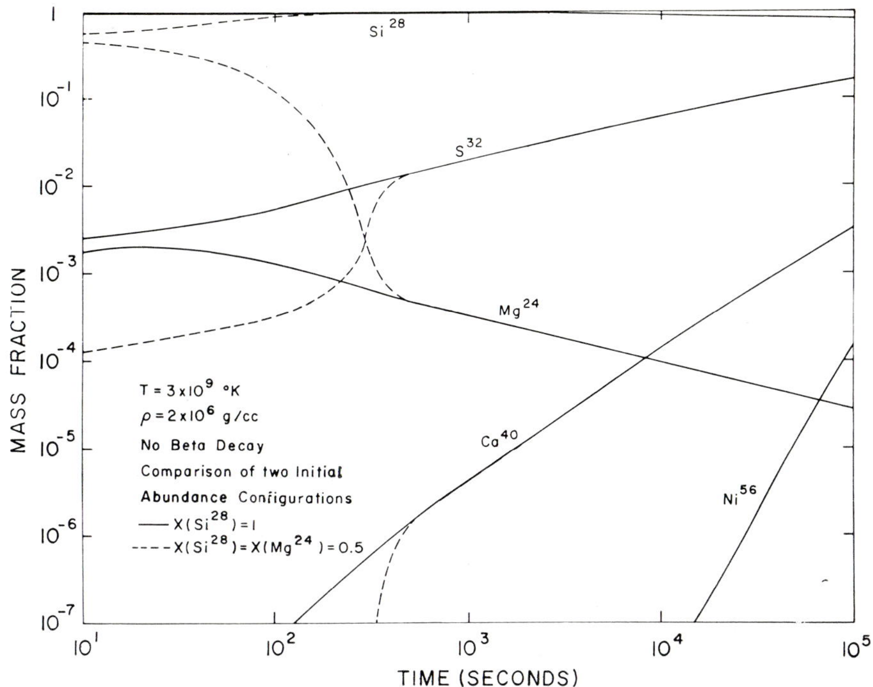

Fig. 6 Comparison between the evolution of abundances for a region composed initially of equal mass fractions of ^{28}Si and ^{24}Mg and that for a pure ^{28}Si concentration.

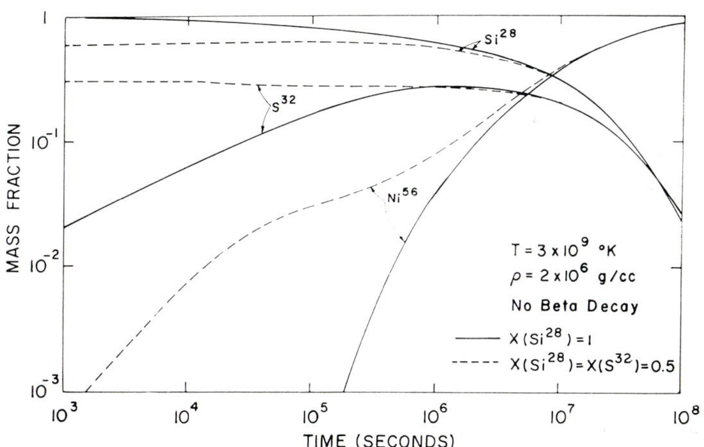

Fig. 7 Comparison between the evolution of abundances for a region composed initially of equal mass fractions of ^{28}Si and ^{32}S and that for a pure ^{28}Si concentration.

(specifically: $t = 4 \times 10^7$ sec. and $\chi(^{28}\text{Si}) = 0.105$). This longer time scale results from the fact that the readjustment of ^{32}S toward equilibrium with ^{28}Si and alpha particles in the early stages does not increase the ^{28}Si abundance as dramatically as in the case of ^{24}Mg. The ^{28}Si mass fraction approaches 0.70, and then begins a long slow decrease with time. For this lower abundance of ^{28}Si, and hence of the lighter elements, fewer alpha particles and protons are made available than at the corresponding time for the pure ^{28}Si evolution.

It is clear from this figure that the abundances of the heavier elements through ^{56}Ni will be greatly enhanced relative to their values for a pure ^{28}Si initial condition for a time of the order of 10^6 seconds. These nuclei are generally characterized by more rapid rates for positron decay and electron capture. This suggests that in any more physical calculation, in which the appropriate weak interaction rates are incorporated, the presence of nuclei heavier than ^{28}Si in the medium at the onset of silicon burning will result in increased neutron enrichment of the resulting abundance distribution. This effect will be lessened considerably at higher temperatures ($T \geq 4 \times 10^9\,^\circ$K) as the time scale for the silicon burning process will be shortened by many orders of magnitude. Furthermore, higher temperature will favor somewhat the destruction of ^{32}S by photodisintegration relative to its destruction by charged particle reactions.

The results presented in this section point out the sensitivity of the abundance patterns realized in equilibrium to the contributions from weak interactions. Some degree of neutron enrichment is required for a region composed initially of ^{24}Mg, ^{28}Si and ^{32}S if a realistic iron group abundance distribution is to result. At higher temperatures, the time scale of the approach to equilibrium is too short to allow any significant neutron enrichment: under these circumstances ^{54}Fe and ^{56}Ni emerge as the most abundant constituents (at high temperatures, higher densities are required to favor ^{56}Ni over its break down to ^{54}Fe $+ 2p$).

The precise temperature-density conditions under which this equilibrium process will take place in stellar interiors are extremely dependent on the detailed stellar models. The preceding discussion would seem to suggest that somewhat lower temperature conditions are required ($T < 4 \times 10^9\,^\circ$K) if this silicon-burning phase is to result in an iron group abundance configuration consistent with the observed solar abundances. The calculations presented here have, however, neglected two important factors: (1) admixtures of neutron rich isotopes in the medium at the onset of silicon burning, and (2) β-decay processes occurring on a longer time scale following the completion of silicon burning. Recent studies of the carbon-burning phase (Arnett and Truran, 1968) based on improved nuclear data (Patterson, Winkler and Zaidins, 1968) reveal that a neutron enrichment of $\sim 1\%$ may result. If this material is then processed through silicon burning at higher temperatures ($T \sim 4$–5 $\times 10^9 \,^\circ$K) a solar-like abundance pattern may emerge. Furthermore, Bodansky *et al.* (1968 a) have pointed out that if account is taken of β-decay processes which occur following the completion of silicon burning, reasonable agreement with the solar abundances can be obtained over a wide range of temperatures. Further investigations of the late stages of evolution are clearly necessary in order to define more precisely the silicon-burning conditions.

III   Thermonuclear Reactions in Supernova Envelopes

The matter in the envelope of a star which is undergoing a supernova explosion will not, in general, be subjected to the very extreme conditions realized in the core. The traversal of these outer regions by an intense shock wave will, however, result in a heating and compression of the medium which can lead to further nuclear processing. In this section the various possible nuclear processes which might take place under these conditions will be considered. While no detailed models of presupernova stars are available, the hydrodynamic studies of Colgate and White (1966) and of Arnett (1966, 1967 a) do give some indication of the range of temperature and density conditions that might be expected.

Consider first the regions immediately surrounding the supernova core. While some degree of compression and heating might have resulted from the infall of this material accompanying the collapse of the core, it will be assumed here that the initial composition has been substantially preserved. This implies that this material should be composed largely of the products of helium, carbon or oxygen burning: nuclei from ^{12}C through ^{28}Si. Some amounts of silicon-burning products may exist in these regions, but much of the material processed through the iron region (if indeed the nuclear transformations have continued this far) must have gone into the formation of the supernova core. The typical temperature and density conditions imparted to these regions by the passage of the shock are $T \gtrsim 5 \times 10^9 \,^\circ$K and $\varrho \gtrsim 10^7$ g/cc.

Truran, Arnett and Cameron (1967) have considered the nuclear transformations which can take place in regions composed of ^{12}C, ^{16}O and ^{28}Si under

these conditions. Specifically, the shock profile determined by Colgate, Grasberger and White (1961) was employed. For this profile, the temperature and
density immediately following the passage of the shock were $T = 5 \times 10^9 \,°K$
and $\varrho = 1.3 \times 10^7$ g/cc. The temperature was found to remain at about
$5 \times 10^9 \,°K$ for approximately 10^{-2} seconds, falling off by an order of mag-

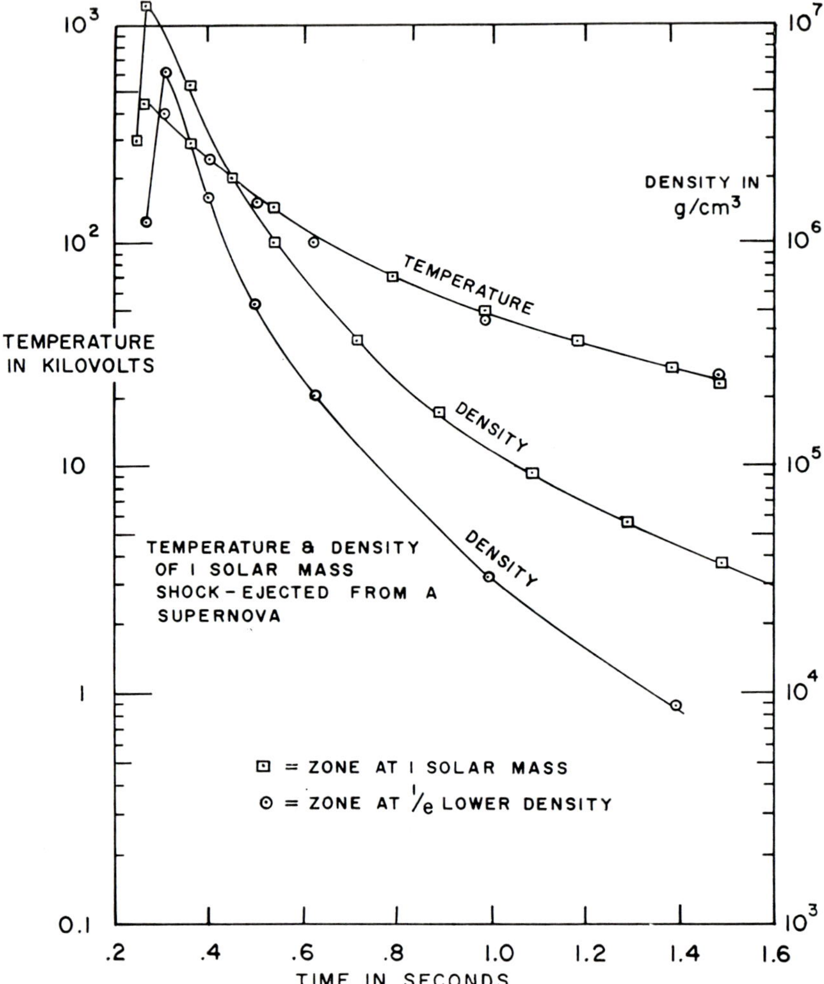

Fig. 8   Shock-wave temperature and density profile (after Colgate *et al.*,
1961). The temperature and density of the shocked stellar material at the
ejection mass cut and at 1/*e* lower density are shown as a function of time.

nitude in the first second as shown in Fig. 8. The nuclear reaction network
defined in the previous section (Fig. 3) was employed for these calculations
as well.

The evolution of the abundances in time for a region composed initially of
pure ^{28}Si is shown in Fig. 9. As the temperature remains rather constant over
the time scale shown in this figure, this calculation is equivalent to an equilibrium calculation at $T = 5 \times 10^9\,°$K. As the time scale is short ($\lesssim 10^{-2}$ sec),

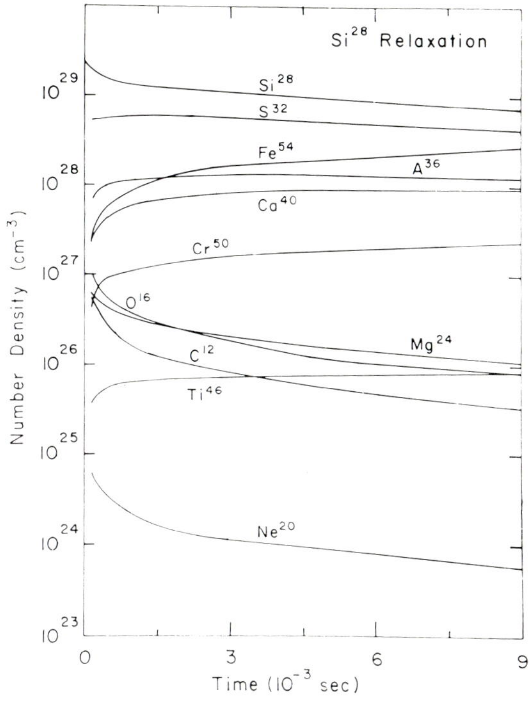

Fig. 9   The evolution of a region composed initially of pure ^{28}Si at
$5 \times 10^9\,°$K.

no neutron enrichment of the material has taken place and the most abundant heavy nuclei produced (^{46}Ti, ^{50}Cr and particularly ^{54}Fe) are very
neutron poor. The free proton number density is also high, the net transformation being represented crudely by 2^{28}Si $\rightarrow$ ^{54}Fe $+ 2p$. (At somewhat
higher densities ($\varrho \gtrsim 10^8$ g/cc) the free protons would be captured by ^{54}Fe
forming ^{56}Ni.)

The implications of these shock conditions for regions composed initially of ^{12}C and ^{16}O have also been considered. The early stage of the evolution for a pure ^{12}C initial composition is shown in Fig. 10. Large abundances of ^{20}Ne, ^{23}Na, ^{24}Mg and ^{27}Al are realized in the early stages due to the importance of the ^{12}C + ^{12}C reaction. The relatively slow effective photodis-

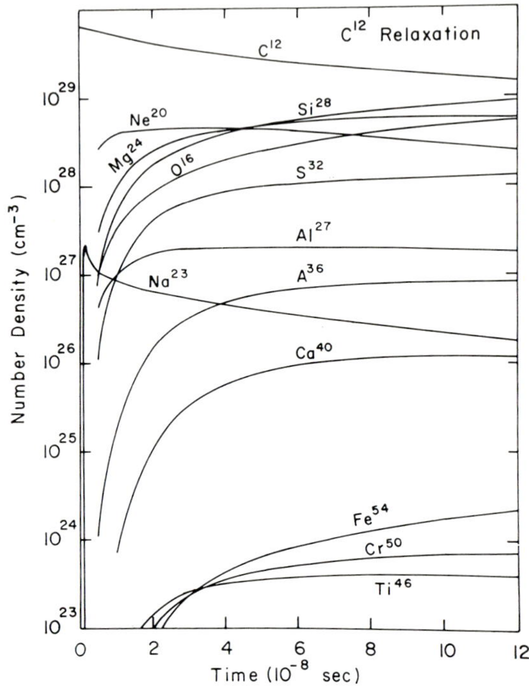

Fig. 10   The early stages of evolution of a region composed initially of pure ^{12}C at 5×10^9 °K.

integration rate of ^{28}Si results in its emergence as the most abundant nuclear species in approximately 5×10^{-6} seconds. The subsequent behavior of this system is controlled by the rate of destruction of silicon. These later stages are shown in Fig. 11. The features of this evolution after $\sim 10^{-5}$ seconds closely resemble those realized for the pure ^{28}Si initial composition at approximately the same time. The subsequent behavior of this system should agree well with that determined for silicon (Fig. 9). The evolution of a region composed initially of pure ^{16}O is characterized by the same general behavior: on a time scale which is short compared to the time required for the production of substantial amounts of iron peak nuclei, ^{28}Si will emerge as the most abundant constituent and its effective rate of destruction will govern the subsequent behavior of the system. Thus, for all three initial compositions, the abundance distribution after $\sim 10^{-3}$ seconds will be well represented as in Fig. 12.

The subsequent evolution of this material is strongly influenced by the expansion and cooling illustrated in the shock profile in Fig. 8. As the tem-

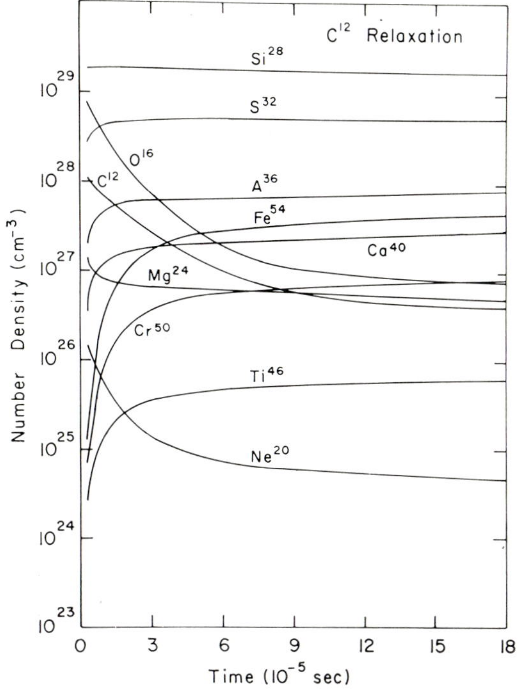

Fig. 11   The later stages of evolution of a region composed initially of pure ^{12}C at 5×10^9 °K.

perature decreases, the photodisintegration rates for protons and alpha particles will be greatly reduced; this results in a favoring of charged particle capture reactions, and the large abundances of free protons and α-particles are exhausted. This behavior is evident in a comparison of the abundance distributions shown in Fig. 12 with that shown in Fig. 13. The abundance distribution at time $t = 0.003$ seconds reflects the high peak temperature of the shock. The free proton and α-particle abundances are extremely high $(n_p \sim 2n(^{54}\text{Fe}))$ and ^{54}Fe is the most abundant heavy nucleus. The temperature at $t = 0.2$ seconds has fallen to $T_9 = 2.6$. The large free proton and α-particle abundances have by this time been substantially depleted, and ^{56}Ni is now the most abundant constituent. In fact, above ^{28}Si the most abundant nuclei through the iron region are all α-particle nuclei. As the proton capture lifetimes for temperatures below $\sim 3 \times 10^9 \,^\circ$K are long compared to the expansion time scale, the freezing of this material is effectively completed. Nuclear positron decays proceeding on a long time scale will result in a readjustment of the nuclei in the iron peak region.

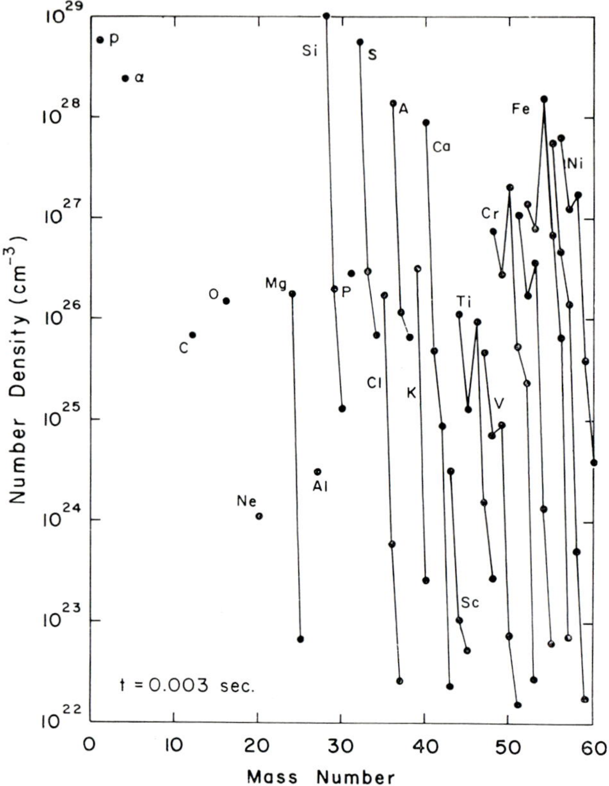

Fig. 12   The abundance distribution after 0.003 seconds ($T \sim 5 \times 10^9 \,^\circ$K).

The large abundance of ^{56}Ni resulting from this freezing process has implications with regard to the observed light curves for supernova explosions. Colgate and McKee (1968) have found that the decay of ^{56}Ni (6.1 days, $E_\beta = 2.11$ MeV) can provide sufficient energy to the expanding supernova shell while the material is opaque to maintain a high temperature and thus to provide a high optical luminosity when the medium becomes optically thin. The lifetime of this decay is roughly consistent with that of the observed peaks in the supernova light curves. The subsequent decay of ^{56}Co (77 days, $E_\beta = 4.57$) may provide a substantial energy contribution to the tail of the light curve.

The calculations of Colgate and McKee require a total mass of the order of $0.20\,M_\odot$ in the form of ^{56}Ni to provide the necessary optical luminosity. In this regard, it is appropriate to consider further the range of conditions for which a substantial ^{56}Ni concentration might result. Preliminary studies of the thermonuclear reaction sequences have been performed for two additional peak shock conditions: (1) $T = 6 \times 10^9\,°K$, $\varrho = 10^7$ g/cc and (2) $T = 7 \times 10^9\,°K$, $\varrho = 10^8$ g/cc. These conditions are consistent with those predicted by the models of Colgate and White (1966) for mass zones near the point where the internal mass fraction $M_r \sim 0.30$. The initial radius is $R_0 \simeq 10^2$ km and the velocity imparted to the medium by the shock is $v \simeq 10^4$ km/sec.

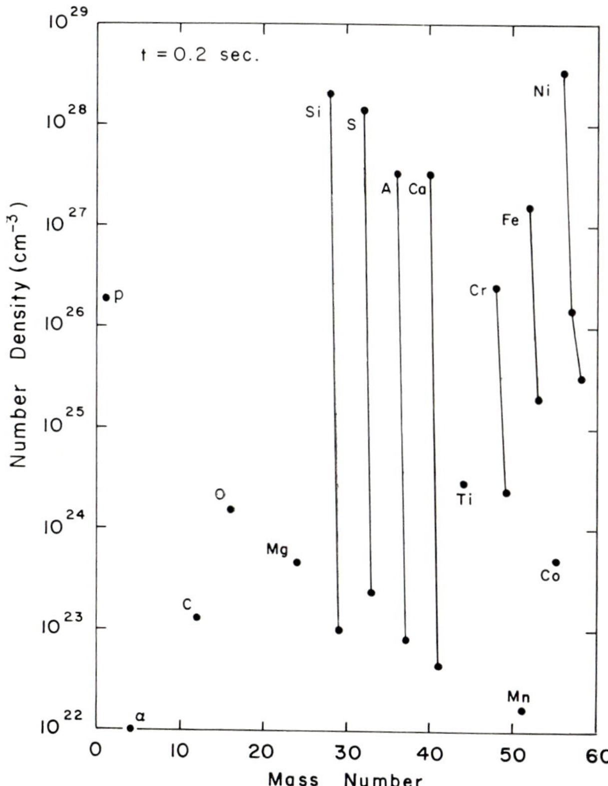

Fig. 13   The abundance distribution after 0.2 seconds ($T \sim 2.6 \times 10^9\,°K$).

These conditions imply a doubling of the radius in a time 10^{-2} seconds. The expansion profile is then given by $\varrho \sim 1/r^3$ and $T \sim 1/r$. The nuclear reaction network described in the previous section was employed, and the initial abundance configuration was assumed to consist of α-particle nuclei in the mass range $12 \leq A \leq 40$.

The results of these calculations are contained in the table. For both temperature-density conditions, the mass fractions of the relevant constituents are presented at two times in the evolution of this material. The "equilibrium" abundances have meaning in that these are realized on a time scale which is extremely short compared to the expansion time scale ($\sim 10^{-2}$), hence the temperature and density are roughly constant and equal to their peak values. The time required for the attainment of these equilibrium configurations are $t \sim 10^{-5}$ seconds for a temperature $T = 6 \times 10^9\,^\circ\text{K}$ and $t \sim 10^{-8}$ seconds for $T = 7 \times 10^9\,^\circ\text{K}$.

For both temperature conditions, the equilibrium configurations are characterized by large number densities of free protons and α-particles and a large abundance of ^{54}Fe. The freezing of this material results in the depletion of the proton abundance and the emergence of ^{56}Ni as the most abundant constituent by mass. A substantial helium abundance survives the freezing process for both of these conditions, though for the higher density case the abundance is considerably lower. These results are preliminary in that the enhancement of the triple-alpha reaction resulting from inelastic scattering with ions and electrons (Shaw and Clayton, 1967) has not been properly taken into account. This would have the effect of further reducing the α-particle abundance resulting from this freezing process. The general conclusion that a substantial mass fraction of ^{56}Ni will be produced under these conditions will not, however, be influenced by these considerations.

The possible role of somewhat weaker shocks in the production of intermediate mass nuclei ($28 \leq A \leq 44$) as well as iron group nuclei is brought out in Fig. 14. This shows the abundance predicted by the calculations performed for a peak temperature $T = 5 \times 10^9\,^\circ\text{K}$ (Fig. 9) at the point when the $^{28}\text{Si}/^{40}\text{Ca}$ ratio has the same value as that in meteorites (Suess and Urey, 1956; Cameron, 1963). Also shown in this figure are the solar system abundances of all stable nuclei in the mass range $24 \leq A \leq 60$. These have been taken from Cameron (1963), except that the iron abundance was reduced by a factor of five from the meteoritic value to be comparable with the present solar value (Aller, 1961). The two distributions have been normalized at ^{28}Si and ^{40}Ca.

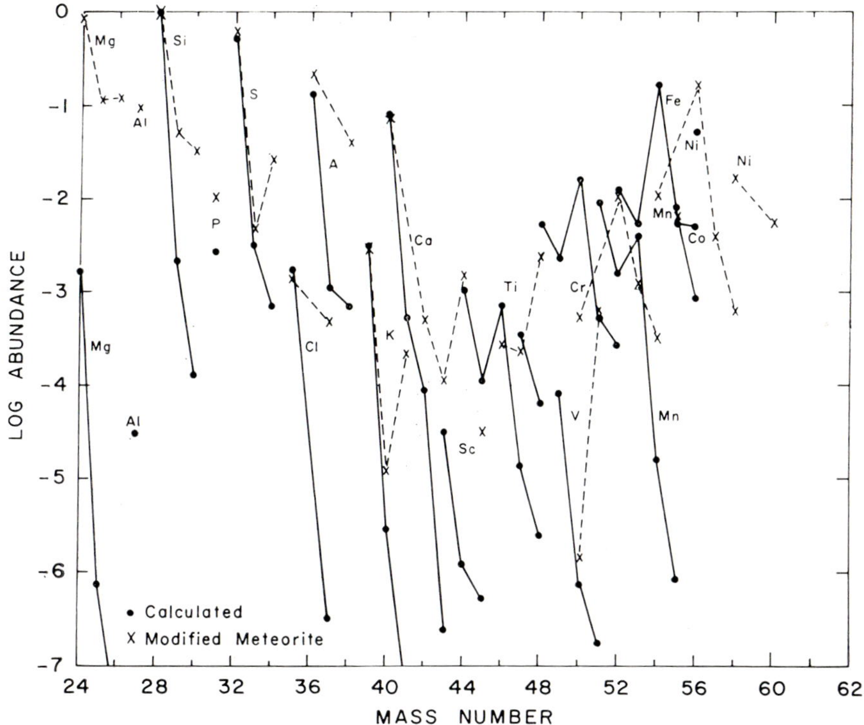

Fig. 14   The abundance distribution at the point at which the ratio (^{28}Si/^{40}Ca) agrees with that observed in the meteorites.

There is a remarkable similarity between the general trends of the abundances in this diagram for nuclei more massive than ^{28}Si. This suggests that the abundances in nature in the element range ^{28}Si-^{40}Ca may be largely produced as a frozen distribution of material incompletely processed toward the iron peak. Detailed agreement is not expected, as some small degree of neutron enrichment in the initial composition is clearly necessary to reproduce the abundances of the minor neutron-excess isotopes. Furthermore, account must be taken both of the freezing of this material and of the β-processes proceeding on a considerably longer time scale. Proceeding in this manner, Bodansky *et al.* (1968a) find that good agreement with various features of the solar system abundance distribution may be obtained for a range of temperatures and densities.

There are two further modes of heavy element synthesis which might proceed under the post-shock conditions in the supernova envelope: rapid neutron capture (the *r*-process) and the process responsible for the production of the bypassed nuclei (the *p*-process). The abundance trends of even mass numbers formed by these two processes and by neutron capture on a slow

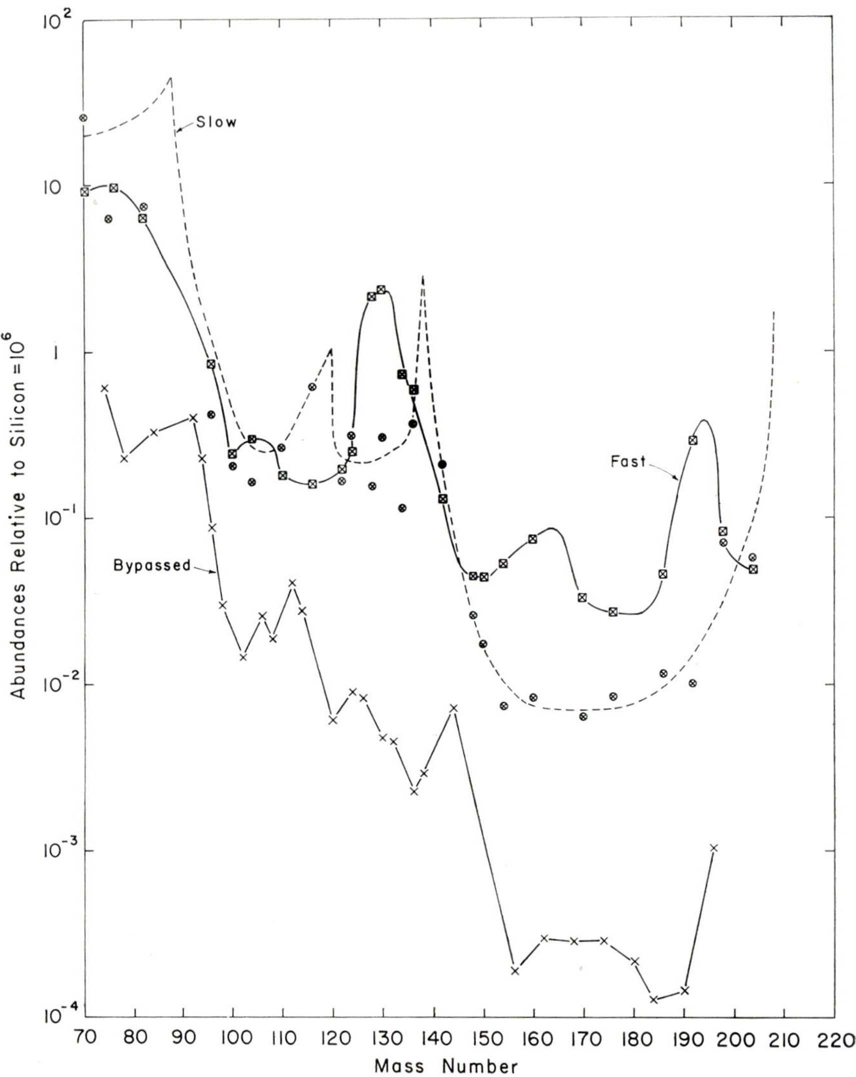

Fig. 15 Trends among heavy nuclei of abundances attributable to neutron capture on fast and slow time scales and to bypassed nuclei (after Cameron, 1967).

time scale (the *s*-process) are shown in Fig. 15 (Cameron, 1967). Neutron capture on a slow time scale is characterized by neutron capture lifetimes which are long compared to typical β-decay lifetimes in the vicinity of the valley of β-stability; the path of this process therefore proceeds along the valley of β-stability, the abundance distribution peaking at the neutron closed shell positions where the neutron capture cross sections decrease. This behavior is evident in Fig. 15 near mass numbers $A = 90$, 138 and 208.

If the neutron capture lifetimes are short compared to the appropriate β-decay lifetimes, the neutron capture path will lie far out on the neutron-rich side of the valley of β-stability. Such a situation might be realized in the presence of an extreme neutron flux; the temperature must be sufficiently low that progress along the neutron-capture chains is not impeded by photodisintegrations. Following the exhaustion of the neutron flux, the capture products approach the valley of β-stability by a series of β-decays. The abundance features near mass numbers $A \sim 130$ and 195 are generally attributed to this neutron-capture process (Seeger *et al.*, 1965). The conditions realized in the cores of supernova are consistent with the requirements for this process, as will be discussed in the next section.

The abundance peak in the vicinity of mass number $A = 103$ and that in the rare earth region (see Fig. 15) are difficult to account for on the basis of the rapid capture process described above. These features may result rather from the exposure of a region containing a few percent by mass of the products of slow neutron capture to a somewhat smaller neutron flux. Truran *et al.* (1967) have suggested that this condition may be realized in the vicinity of a helium-burning shell following the passage of a shock wave through the supernova envelope. Neutrons released in (α, n) reactions proceeding on ^{25}Mg, ^{26}Mg and ^{29}Si, for a shock of peak temperature $T \sim 3 \times 10^9 \, ^\circ$K, resulted in a neutron number density exceeding 2×10^{18} cm^{-3} for a time of the order of 10^{-2} seconds. Detailed calculations of the subsequent neutron-capture process were carried out and found to be very sensitive to the neutron-capture cross-section estimates for the neutron-rich isotopes. Further improvements in the cross-section calculations are necessary before any definite conclusions can be drawn from this work.

The difficulties encountered above would be readily overcome in the presence of a somewhat larger neutron flux, $n_n \lesssim 10^{20}$ cm^{-3}. This flux is still far less than that required to account for the dominant rapid capture features (Seeger *et al.*, 1965). Schwarzschild and Härm (1967) have recently carried numerical calculations through the first 4 million years of shell helium burning for a Population II star of 1.2 solar masses. The thermal instability associated with this shell-burning phase results in a sequence of "relaxation oscillations" in which a growing convective zone ultimately mixes protons from the hydrogen-rich layers above with the products of helium burning. Sanders (1967) has found that for relatively small admixtures of protons into a carbon (oxygen) region, the major product will be ^{13}C (^{17}O). For both of these cases (α, n) reactions proceeding on a somewhat longer time scale may release sufficient neutrons to sustain a slow neutron-capture process.

These conditions are promising as well for the rapid-capture process considered above (Truran *et al.*, 1968). The mixing of protons into a carbon and oxygen region will result in the production of ^{14}N as well as ^{13}C and ^{17}O; in fact, for large proton admixtures the abundance of ^{14}N should exceed those of ^{13}C and ^{17}O. In the presence of an appreciable neutron abundance this ^{14}N can be destroyed by

$$^{14}N(\alpha,\gamma)^{18}F(\beta^+\nu)^{18}O(\alpha,\gamma)^{22}Ne$$

The subsequent passage of a shock wave through such a region can then result in the production of a more substantial neutron flux ($\sim 10^{20}$ cm^{-3}) by

$$^{22}Ne(\alpha,n)^{25}Mg(\alpha,n)^{28}Si$$

This is a more promising model than the helium-shell model considered previously, but no detailed calculations have as yet been carried out.

The production of the bypassed nuclei may also take place in the supernova envelope in the wake of the shock. The bypassed nuclei are nuclei on the neutron-deficient side of the valley of β-stability which cannot be produced by any neutron-capture process. In principle, these can be formed by neutron photodisintegration or proton-capture reactions (the *p*-process) proceeding on the products of earlier neutron-capture synthesis.

The abundance variations of the bypassed nuclei and those of the low neutron-capture products show a striking similarity as is evident in Fig. 15. The principle differences are that the abundance ratio of bypassed nuclei to slow capture products decreases with increasing mass number and that the abundance level for bypassed nuclei remains rather constant right up to the positions of the neutron closed shells beyond which it falls precipitously. These are precisely the features one would predict if the products of slow neutron capture were raised to a temperature of $T \sim 2$–3×10^9 °K for a short period in a very proton-rich medium. These conditions are consistent with the traversal of a supernova shock wave through the outer regions (of approximately solar composition) of a supernova envelope.

IV   Element Synthesis in Supernova Cores

The most extreme conditions to which matter will be subjected in supernovae
are realized in the supernova core. Recent studies of the dynamics of the cores
of highly evolved massive stars (Colgate and White, 1966; Arnett, 1966,
1967a) provide the best estimates of these physical conditions. In both investigations it was found that a considerable fraction of the core mass will be
ejected. It is of interest to consider the implications of these models for heavy
element synthesis.

At the extreme temperature and density conditions realized in the core the
matter is composed predominantly of neutrons. As the ejected matter expands and cools, these neutrons will be converted to protons, helium and
heavier elements. Arnett and Cameron (1967) found that the weak interactions should freeze out in the early stages of expansion of the ejected material at a temperature of approximately 20 billion degrees and a density of
7×10^9 g/cc. The neutron to proton ratio realized in this freezing process
was approximately 8; this ratio was found not to be a sensitive function of
the freezing density. For this neutron to proton ratio only a fraction of the
mass will be converted to helium. The short time scale of the ensuing expansion ($\lesssim 10^{-1}$ seconds) implies that a considerable abundance of free neutrons
will survive the charged particle reaction stages.

Truran, Arnett, Tsuruta and Cameron (1968) have investigated the subsequent thermonuclear evolution of this material. The temperature-density
history of the medium employed in these calculations is shown in Fig. 16.
The initial temperature considered here is $10^{10}\,°K$, at which point the weak
interactions have frozen out and the formation of helium from neutrons and
protons will proceed rapidly. As uncertainties in the hydrodynamic calculations can readily distort the expansion adiabat, calculations were performed
both for the density profile shown in the figure and for an adiabat at a factor
of 10 lower density. Arnett and Cameron (1967) have shown that substantial
amounts of ^{12}C are produced in equilibrium with neutrons, protons and
α-particles during the expansion of the neutron-rich material along these
adiabats. The question then is whether a true nuclear statistical equilibrium
will be approached and maintained during the expansion.

The nuclear reaction network employed in these calculations is shown in
Fig. 17. In the light element region the more important reactions linking
neutrons, protons and α-particles are included. The rates for these reactions

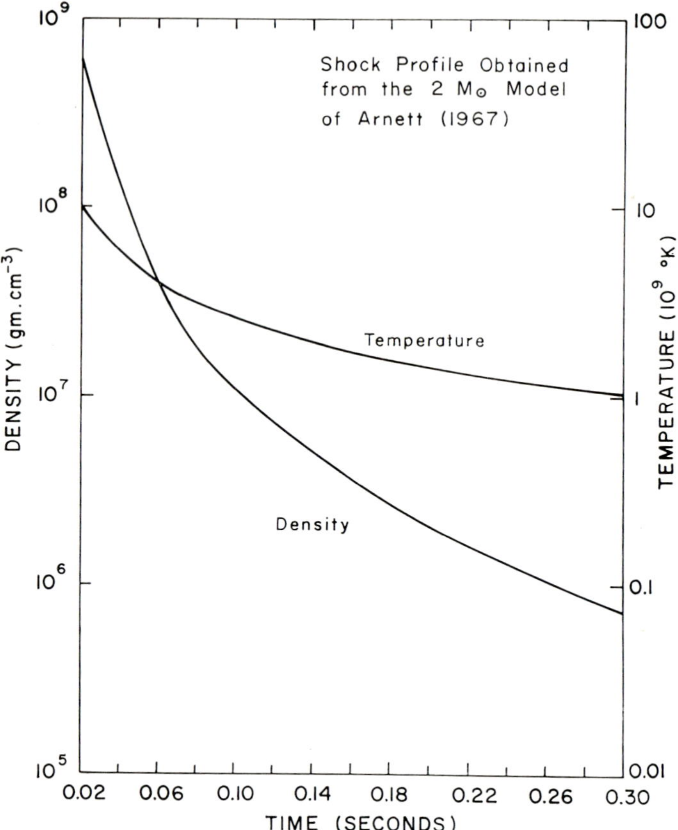

Fig. 16 Shock-wave temperature and density profile employed for
studies of element synthesis in supernova cores.

were taken from Wagoner *et al.* (1967). For the heavy elements, a sequence of
α-particle and neutron-capture reactions was defined proceeding from ^{12}C
to ^{86}Kr. Experimental determinations of the reaction rates were employed
for the triple-α reaction and for the α-capture reactions on ^{12}C, ^{16}O, ^{20}Ne
and ^{24}Mg (Truran *et al.*, 1966a,b; Reeves, 1965; Fowler *et al.*, 1967). As ^{12}C
is found to be in equilibrium with α-particles throughout the buildup process, a consideration of the enhancement of the triple-alpha rate by inelastic
scattering with electrons, ions, and neutrons (Shaw and Clayton, 1967) is not
relevant.

The nuclear-reaction pathway chosen here lies close to the ordinary valley of β-stability, whereas that pathway along which the approach to equilibrium will take place under the conditions considered here ($n_n/n_p = 8$) presumably lies in the region of very neutron-rich heavy nuclei. The reason for this choice was that the techniques used to determine the appropriate nuclear reaction cross sections (Truran *et al.*, 1966a) will not give reliable predictions for very neutron-rich nuclei. The point of view adopted here was that if nuclear statistical equilibrium can be approached by *any* set of reaction pathways during the expansion along the adiabat, then it should certainly be attained by the *fastest* pathway.

The mass fractions of the more important nuclear constituents are plotted as a function of time for the high-density profile in Fig. 18. The initial composition was taken to be neutrons and protons in the ratio 8 : 1. It is evident from the behavior in the early stages that an equilibrium has been rapidly established among neutrons, protons, α-particles and ^{12}C. As the tempera-

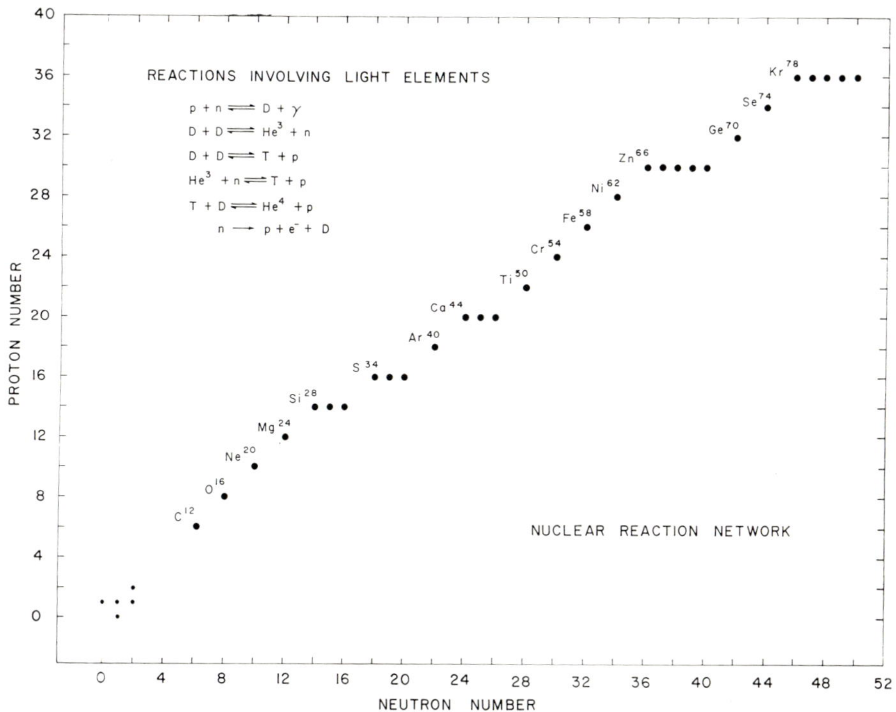

Fig. 17 Schematic of the nuclear reaction network employed for studies of element synthesis in supernova cores.

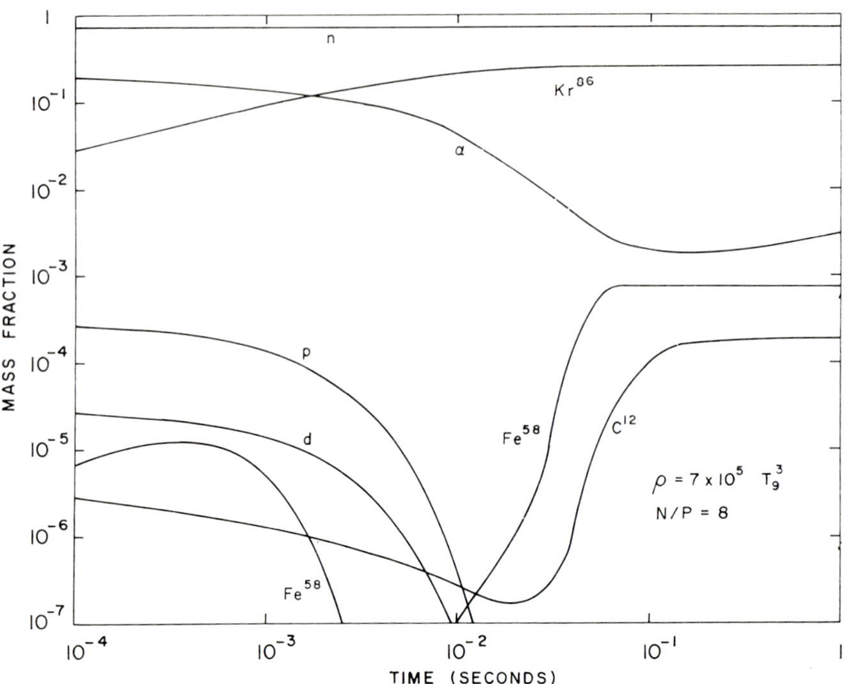

Fig. 18 The evolution of a region composed initially of neutrons and protons in the ratio $N/P = 8$ for the high density adiabatic cooling curve.

tures decreases, and the α-particle buildup past ^{12}C becomes more rapid, the abundances of the light elements and ^{12}C decrease while the abundances of intermediate nuclei and ultimately ^{86}Kr are rapidly increasing. At a temperature of approximately 4 billion degrees (approximately 2×10^{-2} sec) the reactions involving the heavy elements freeze out in the sense that the nuclear-reaction time scale becomes long compared to the expansion time scale. The final abundance of ^{86}Kr is 20% by mass.

The results of a similar calculation performed for the lower-density profile are shown in Fig. 19. Here the buildup of ^{12}C is somewhat impeded by the lower density. However, the final ^{86}Kr abundance still approaches 10% by mass.

The effects of varying the total neutron to proton ratio realized in freezing are shown in Fig. 20. Here the ratio $n_n/n_p = 3$, and the low-density adiabat is employed. Generally, the protons will be converted completely to helium,

hence a larger fraction of the mass can be converted to heavy elements. The
^{86}Kr abundance realized in this calculation approaches 50% by mass.

These results suggest that the breakthrough past ^{12}C to heavy elements
can take place readily under these conditions. However, the simple linear
chain past carbon does not provide an accurate representation of the flows
through the iron peak and beyond, nor does it provide a realistic abundance
configuration. The final distribution of nuclei resulting from this freezing
process may be estimated on the assumption that the nuclear reactions are
sufficiently rapid that a true condition of nuclear s^{t}atistical equilibrium is
attained (Hoyle, 1946; Tsuruta and Cameron, 1965).

The equilibrium abundances obtained for a temperature of 4×10^9 °K, a
density of 4.48×10^7 g/cc and a total neutron to proton ratio of 8 are shown
in Fig. 21. These conditions are those predicted for the high density adiabat
at the point at which the charged particle reactions have frozen out. The
masses employed for nuclei in the neutron-rich regions far from the valley

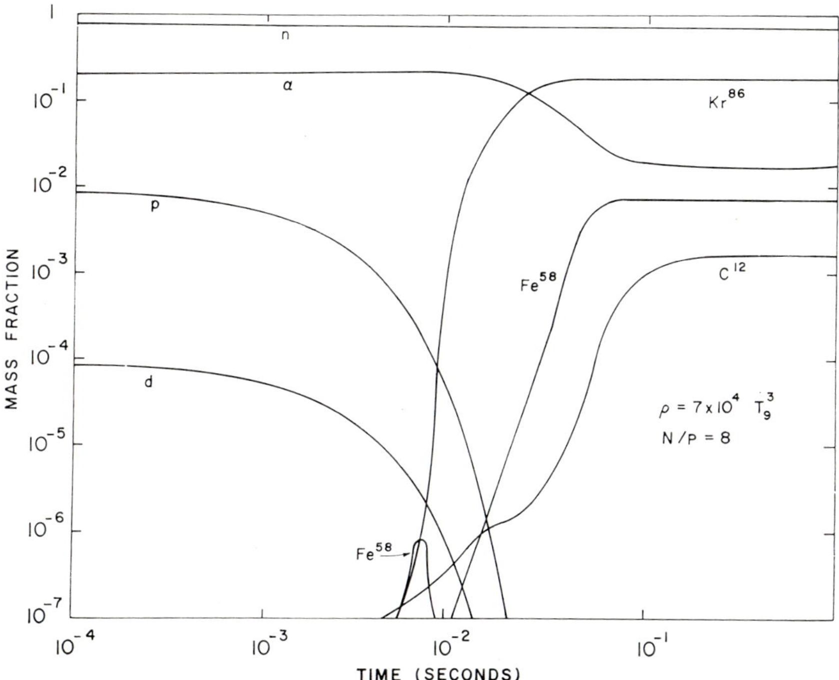

Fig. 19 The evolution of a region composed initially of neutrons and
protons in the ratio $N/P = 8$ for the low density adiabatic cooling curve.

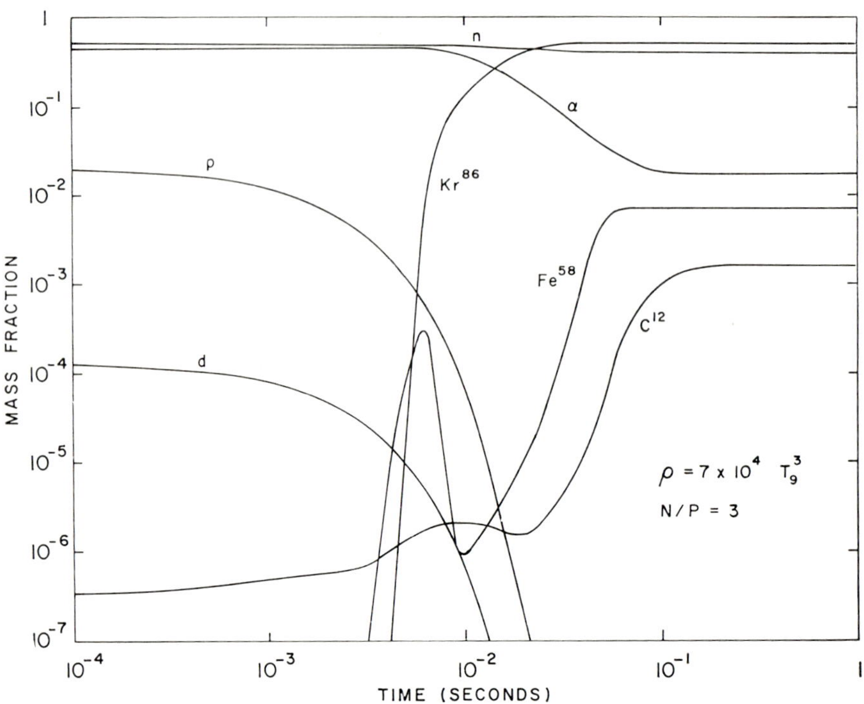

Fig. 20   The evolution of a region composed initially of neutrons and protons in the ratio $N/P = 3$ for the low density adiabatic cooling curve.

of β-stability are those predicted by the exponential mass formula of Cameron and Elkin (1965).

The peak nucleus in equilibrium under these conditions is ^{78}Ni, having 28 protons and 50 neutrons. It is understandable that this nucleus, possessing "magic" numbers of both protons and neutrons, is exceptionally stable. However, at slightly lower temperatures the peak tends to shift toward heavier nuclei. The equilibrium abundance distribution predicted for the same adiabat at a temperature of $3 \times 10^9\,^\circ$K is found to peak at ^{120}Sr (38 protons and 82 neutrons); this abundance peak is one neutron closed shell removed from the valley of β-stability.

The equilibrium abundances for the low-density adiabat at a freezing temperature of $4 \times 10^9\,^\circ$K is shown in Fig. 22. The total neutron to proton ratio is again taken to be 8. The abundance peak here is again in the vicinity of ^{78}Ni, and remains so for temperatures down to $3 \times 10^9\,^\circ$K.

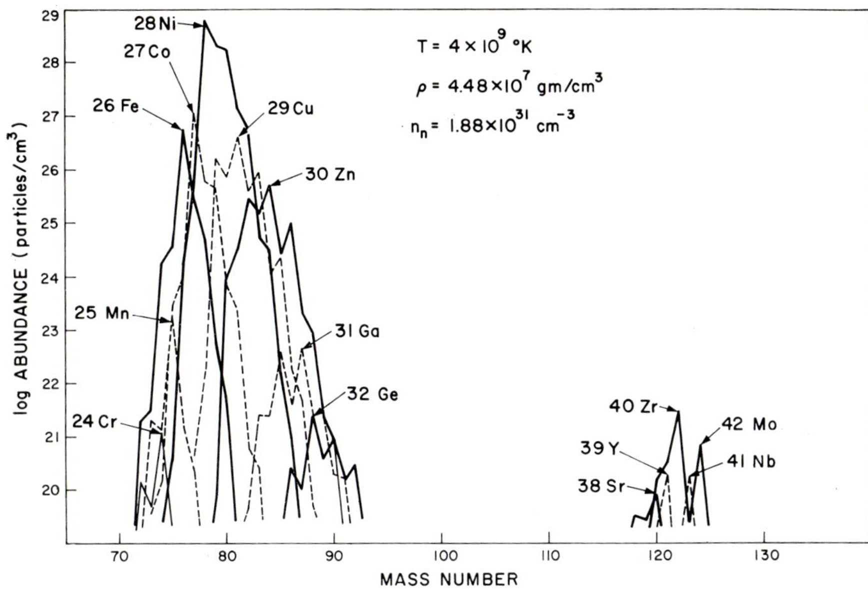

Fig. 21   The equilibrium abundances of nuclei at the freezing temperature ($T = 4 \times 10^9$ °K) for the high density adiabatic cooling curve.

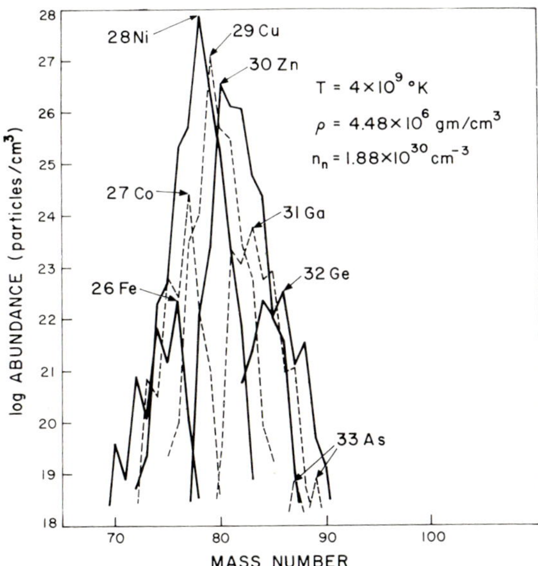

Fig. 22   The equilibrium abundances of nuclei at the freezing temperature ($T = 4 \times 10^9$ °K) for the low density adiabatic cooling curve.

For both adiabats, the abundance peaks predicted for a freezing temperature of $4 \times 10^9\,°$K fall in the vicinity of mass number $A = 78$. As a reaction pathway dominated by α-particle capture reactions along the valley of β-stability has been sufficient to produce a substantial abundance at mass $A = 86$, it is reasonable to assume that the time scale of the expansion is sufficient to allow a true nuclear statistical equilibrium to be attained. It is therefore appropriate to consider the implications of these supernova shock conditions with regard to heavy element synthesis.

While the charged-particle reactions are no longer contributing to the buildup of heavy elements below $4 \times 10^9\,°$K, and progress toward elements of higher Z is therefore impeded, neutron-capture buildup under these conditions should be limited only by neutron-photodisintegration reactions. For the equilibrium abundance distribution displayed in Fig. 22, the neutron number density is 1.88×10^{30} cm^{-3} and the ratio of neutrons to heavy nuclei is approximately 200. As the temperature is further decreased, the neutron photodisintegration rates will decrease, allowing successive neutron captures to proceed toward nuclei which are increasingly unstable.

Seeger, Clayton and Fowler (1965) have pointed out that if the observed abundance peaks at $A = 130$ and 195 are to be understood in terms of nuclear shell structure, an environment characterized by a high temperature and neutron number density ($T \sim 10^9\,°$K, $n_n \simeq 10^{24}$ cm^{-3}) is required. These conditions are necessary, on their model, for the establishment of a thermal equilibrium between the isotopes of any element, neutrons and protons on a time scale which is short compared to typical β-decay lifetimes. It is clear, however, that these conditions are extremely sensitive to mass formula predictions of the neutron binding energies and β-decay energies of the neutron-rich isotopes formed in the rapid neutron-capture process. Truran *et al.* (1968) have argued that the rapid neutron-capture buildup can be terminated either by the exhaustion of the neutron flux or by the expansion of the medium to extremely low densities. In order to determine which of these is the limiting consideration, and to calculate the final abundances resulting from this neutron-capture process, the subsequent history of the medium must be followed in detail. Further hydrodynamic studies of the supernova mechanism are required for this purpose.

V   Discussion

The calculations described in this paper have demonstrated that the conditions predicted by current hydrodynamic studies of the supernova mechanism are consistent with the production of heavy elements in the expanding shells of supernova. The relative contributions of the various processes considered here are uncertain; it should be noted, however, that the products of the rapid neutron-capture process have an observed abundance relative to the silicon to iron region in the range 10^{-4}–10^{-6}. If supernova explosions are to account for the abundances both of the iron group and of the very heavy elements, it is clear that only a small part of the mass can be processed by rapid neutron capture in the manner outlined in section IV. Further refinements both in the stellar evolutionary models for the late stages of evolution and in the hydrodynamic models of supernova explosions, as well as improvements in estimates of the relevant nuclear parameters, are required before any definite statement can be made regarding the extent to which these thermonuclear processes will proceed. It seems likely, however, that at least some reasonable fraction of the heavy element abundances observed in the solar system has resulted from such events.

Clayton, Colgate and Fishman (1968) have discussed a possible test of one aspect of these calculations. Assuming that ^{56}Ni is the most abundant nucleus resulting from silicon burning at typical shock temperatures, they find that

Table I   Mass fractions resulting from thermonuclear processes
in high temperature shocks

| Nucleus | $T_9 = 6,\ \varrho = 10^7$ | | $T_9 = 7,\ \varrho = 10^8$ | |
	Equilibrium Mass Fraction $t \lesssim 10^{-5}$ sec.	Frozen Mass Fraction $t \sim 10^{-2}$ sec.	Equilibrium Mass Fraction $t \lesssim 10^{-8}$ sec.	Frozen Mass Fraction $t \sim 10^{-2}$ sec.
n	3.4×10^{-5}	4×10^{-23}	7.6×10^{-5}	2×10^{-14}
p	0.028	7×10^{-4}	0.028	2.0×10^{-3}
α	0.37	0.29	0.37	0.056
^{28}Si	6.5×10^{-3}	4.1×10^{-5}	3.8×10^{-3}	8.2×10^{-6}
^{52}Fe	6.5×10^{-4}	1.6×10^{-3}	1.3×10^{-3}	5.1×10^{-4}
^{54}Fe	0.17	3×10^{-14}	0.14	2.7×10^{-15}
^{56}Fe	0.064	1×10^{-23}	5.4×10^{-3}	1.1×10^{-20}
^{56}Ni	2.6×10^{-3}	0.66	3.9×10^{-3}	0.87
^{58}Ni	0.039	1.8×10^{-6}	0.042	0.042

the mass of ^{56}Ni (0.14 M$_\odot$) required to explain the optical luminosity for Type I supernovae (Colgate and McKee, 1968) should give rise to gamma-ray lines from the subsequent decay of ^{56}C (77 days) which might be detectable in young supernova remnants. Furthermore, gamma-rays resulting from the decay of ^{44}Ti, which is produced under the same neutron poor conditions as is ^{56}Ni, should be detectable on a somewhat longer time scale. A positive identification of any of these lines in supernova remnants would strongly support many aspects of the recent investigations of the nature of supernova explosions.

References

Aller, L.H., 1961, *The Abundances of the Elements*, Interscience Publishers, New York.
Alpher, R.A., and Herman, R.C., 1950, *Rev. Mod. Phys.* **22,** 153.
Arnett, W.D., 1966, *Can. J. Phys.* **44,** 2553.
Arnett, W.D., 1967a, *Can. J. Phys.* **45,** 1621.
Arnett, W.D., 1967b, *On Supernova Hydrodynamics*, preprint.
Arnett, W.D., and Cameron, A.G.W., 1967, *Can. J. Phys.* **45,** 2953.
Arnett, W.D., and Truran, J.W., 1968, *Carbon-Burning Nucleosynthesis at Constant Temperature*, preprint.
Bodansky, D., Clayton, D.D., and Fowler, W.A., 1968a, *Phys. Rev. Lett.* **20,** 161.
Bodansky, D., Clayton, D.D., and Fowler, W.A., 1968b, *Nuclear Quasi-Equilibrium During Silicon Burning*, preprint.
Burbidge, E.M., Burbidge, G.R., Fowler, W.A., and Hoyle, F., 1957, *Rev. Mod. Phys.* **29,** 547.
Cameron, A.G.W., 1957, Chalk River Report CRL-41.
– 1959a, *Astrophys. J.* **130,** 429.
– 1959b, *Astrophys. J.* **130,** 895.
– 1963, *Nuclear Astrophysics*, Yale University lectures.
Cameron, A.G.W., and Elkin, R.M., 1965, *Can. J. Phys.* **43,** 1288.
Chiu, H.Y., 1966, *Presupernova Evolution*, in *Stellar Evolution* (ed. by R.F.Stein and A.G.W.Cameron), Plenum Press, New York.
Clayton, D.D., Colgate, S.A., and Fishman, G.J.: 1968, *Gamma-Ray Lines from Young Supernova Remnants*, preprint.
Colgate, S.A., Grasberger, W.H., and White, R.H., 1961, Lawrence Radiation Lab. UCRL-6471.
Colgate, S.A., and McKee, C., 1968, *Supernova Luminosity*, to be published.
Colgate, S.A., and White, R.H., 1966, *Astrophys. J.* **143,** 626.
Fowler, W.A., Caughlan, G.R., and Zimmerman, B.A., 1967, *Ann. Rev. Astron. Astrophys.* **5,** 525.
Fowler, W.A., and Hoyle, F., 1964, *Astrophys. J. Suppl.* **91.**
Greenstein, G.S., 1968, Yale University thesis.
Hansen, C.J., 1966, Yale University thesis.

Hoyle, F., 1946, *Monthly Notices Roy. Astron. Soc.* **106**, 23.

Patterson, J.R., Winkler, H., and Zaidins, C.S., 1969, *Experimental Investigation of the Stellar Nuclear Reaction* $^{12}C + {}^{12}C$ *at Low Energies*, preprint.

Peebles, P.J.E., 1966, *Phys. Rev. Lett.* **16**, 410.

Penzias, A.A., and Wilson, R.W., 1965, *Astrophys. J.* **142**, 419.

Reeves, H., 1965, *Stellar Structure*, Chap. 2 (ed. by L.H. Aller and D.B. McLaughlin), University of Chicago Press, Chicago.

Reeves, H. and Salpeter, E.E., 1959, *Phys. Rev.* **116**, 1505.

Sanders, R.H., 1967, *S-Process Nucleosynthesis in Thermal Relaxation Cycles*, preprint.

Schwarzschild, M., and Härm, R., 1967, *Hydrogen Mixing by Helium Shell Flashes*, preprint.

Seeger, P.A., Fowler, W.A., and Clayton, D.D., 1965, *Astrophys. J.* Suppl. XI, No. **97**.

Shaw, P.B., and Clayton, D.D., 1967, *Particle-Induced Electromagnetic De-excitation of Nuclei in Stellar Matter*, preprint.

Stothers, R., and Chiu, H.Y., 1962, *Astrophys. J.* **135**, 963.

Suess, H.E., and Urey, H.C., 1956, *Rev. Mod. Phys.* **28**, 53.

Truran, J.W., Arnett, W.D., and Cameron, A.G.W., 1967, *Can. J. Phys.* **45**, 2315.

Truran, J.W., Arnett, W.D., Tsuruta, S., and Cameron, A.G.W., *Astrophys. Space Sci.* **1**, 129.

Truran, J.W., Hansen, C.J., Cameron, A.G.W., and Gilbert, A., 1966a, *Can. J. Phys.* **44**, 151.

Truran, J.W., Cameron, A.G.W., and Gilbert, A., 1966b, *Can. J. Phys.* **44**, 563.

Truran, J.W., 1968a, *A Numerical Test of the Quasi-Equilibrium Approximation for Stellar Silicon Burning*, preprint.

Truran, J.W., 1968b, *The Influence of a Variable Initial Composition on Stellar Silicon Burning*, preprint.

Tsuda, H., 1963, *Progr. Theoret. Phys.* **29**, 29.

Tsuruta, S., and Cameron, A.G.W., 1965, *Can. J. Phys.* **43**, 2056.

Wagoner, R.V., Fowler, W.A., and Hoyle, F., 1967, *Astrophys. J.* **148**, 3.

8

PROPERTIES OF NEUTRON STARS

A. G. W. Cameron

Belfer Graduate School of Science
Yeshiva University
New York, N. Y.

THE PROBLEM of neutron stars first arose in the 1930's and was a topic of considerable theoretical interest at the time. The first computations relating to these objects were performed by Landau and Oppenheimer, among others; at that time Zwicky suggested that a neutron star might be the remnant of a supernova. However, no neutron stars were discovered, so that they remained as theoretical curiosities. On the other hand, calculations pertaining to gravitational collapse indicate that objects of nuclear density will be formed as supernova remnants, and whether or not they will go into continued collapse can only be determined if we know the properties of a stable neutron star.

If a neutron star is in fact formed as a remnant of a supernova explosion, the environment in which it is formed is dynamically very violent, and possibilities exist for storing energy in various forms. This is particularly interesting in view of the discovery of x-ray sources which might directly or indirectly be associated with the presence of neutron stars in the remnants of supernova explosions. If we consider the mechanisms for energy storage in a neutron star which should be investigated, we find four possible energy stores: thermal, rotational, vibrational, and magnetic.

Before we consider these mechanisms in detail, we must first discuss the basic properties of a neutron star. Consider a static, cold neutron star model. Here we have to solve the general relativistic equations of hydrostatic equilibrium; however, it turns out that the actual departures from the Newtonian picture are not very great and do not introduce any fundamental difference in the characteristics of neutron stars until the upper limit for the stable mass of such objects is approached. There are some additional complications in

147

determining the structure of a neutron star as compared to the problems that arise in calculating the structure of a white dwarf. In each case we are dealing with a degenerate gas, but in the case of the white dwarf the pressure is due to the electrons and the equation of state is modified only slightly by the

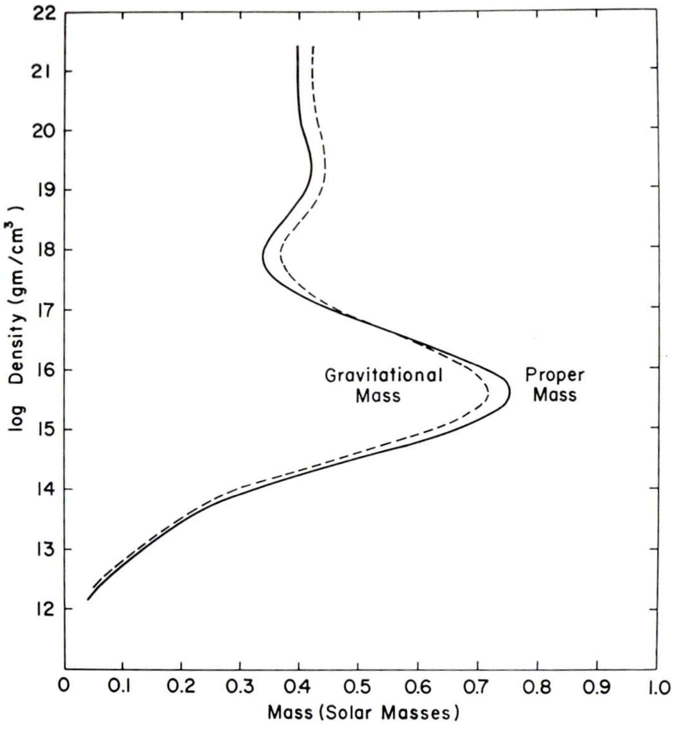

Fig. 1   Relation between central density and mass for neutron stars composed of noninteracting neutrons.

interactions between the particles. In the case of the neutron star this is no longer true, and one must try to make some modification of the equation of state to reflect these interactions. Figure 1 shows the relationship between the central density and mass of a neutron star in which the equation of state corresponds to noninteracting neutrons. One observes an increase in the stable mass as the central density increases until an upper limit to the stable mass is reached at a density of about 10^{15} gm/cm^3. Beyond this peak a general relativistic instability sets in which assures that the star will be unstable and will collapse.

Various modifications of the equation of state have been made in order to approach more realistic conditions. The most important ones are shown in Fig. 2. Salpeter's equation of state is very soft in the low-density region and is quite different from the others in that respect. It was derived from the pro-

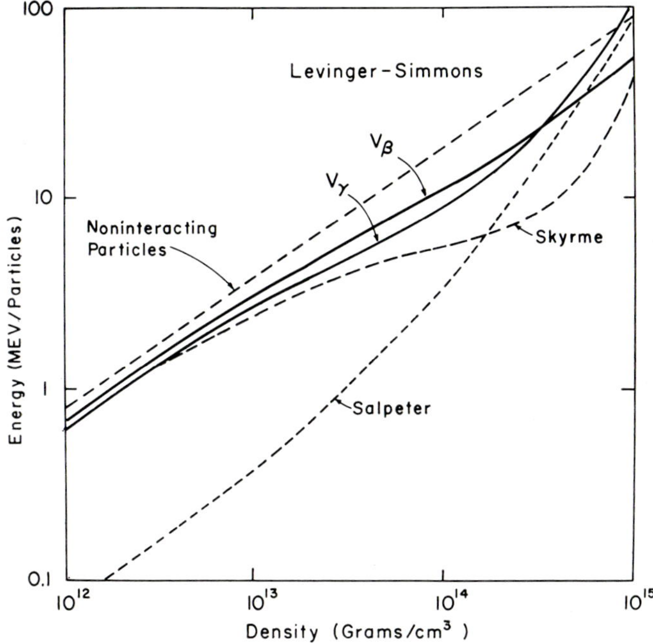

Fig. 2   Various equations of state which have been used in the construction of neutron stars.

perties of nuclei and may be a better equation in the sense that none of the others takes into account the probability that for relatively low densities there will be a clustering of nucleons to form quasi-nuclei. This effect will increase the proton-to-neutron ratio and very greatly soften the equation of state. Such a modification has yet to be made in the other equations, so that considerations based on Salpeter's equation are interesting in view of that possibility.

Several neutron star models were computed by Tsuruta based on two equations of state, referred to as V_β and V_γ, which were suggested byLevinger and Simmons for a neutron gas. They are similar in the low-density region but differ markedly at high densities in that the V_γ equation becomes stiff much more rapidly than the V_β. Some recent work has been done by Weiss in applying these equations to the study of nuclear matter; he finds that the V_β potential gives very poor results for nuclear matter whereas the V_γ potential gives fairly good results. On these grounds it appears that the V_γ equation is much the preferable one to use.

An additional consideration with respect to the equation of state is that at very high densities the pressure must not become so high that the sound speed exceeds the speed of light. In fact for a reasonably-behaved fluid the sound speed should not exceed $c/\sqrt{3}$. Therefore one must cut off the equation of state in the very high density limit to prevent such a situation. Another feature at very high densities is that the Fermi levels of the particles become very high and it becomes energetically favorable to convert some of the neutrons and protons into various baryons. Figure 3 shows the results of

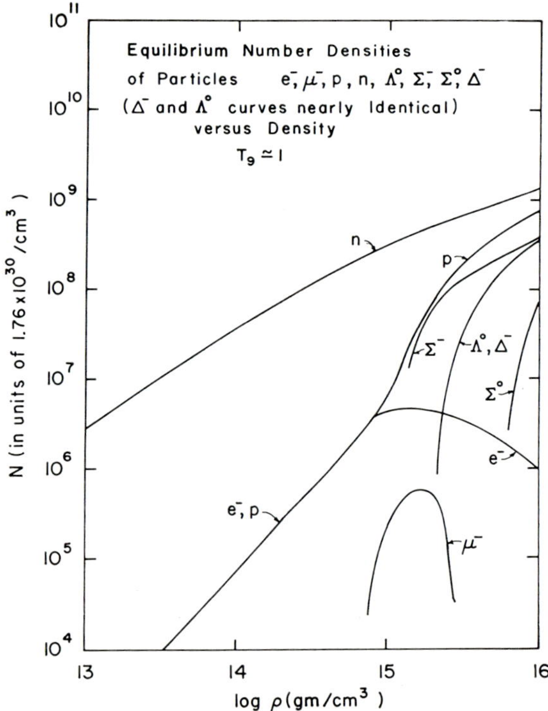

Fig. 3 Composition of cool hyperonic matter at high density. No corrections have been made for any alteration in the effective masses of the particles.

calculations by Hansen of the relative abundances. His calculations neglect the modifications of the effective masses of the baryons which would result from the potentials between the baryons and the neutrons and protons. When the corrections for the effective masses are made, the proton abundance will increase by about a factor of 3, as shown by Wolf. Initially, electrons and protons have the same abundance, but as the electron Fermi level

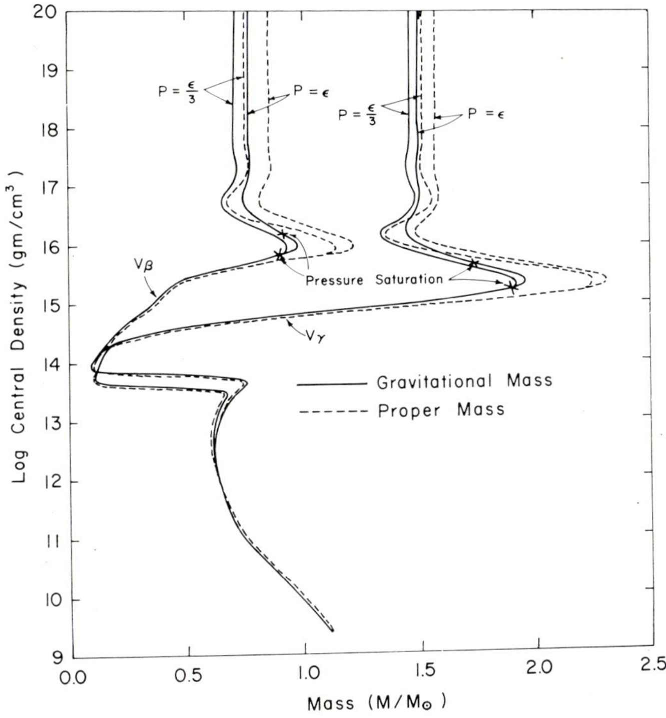

Fig. 4 Relationship between central density and mass for neutron stars based on the Levinger-Simmons potentials and with ionic surface layers.

becomes large it becomes energetically favorable to have some muons present. In the region above 10^{15} gm/cm^3, which is just over twice nuclear density, Σ^-, Λ^0 and Δ^- particles start to appear.

Figure 4 shows the central density-mass relationship corresponding to a composite equation of state which varies from ordinary ions on the surface of the star to various mixtures of neutrons, protons and baryons in the interior. Both the V_β and V_γ potentials give curves which are qualitatively

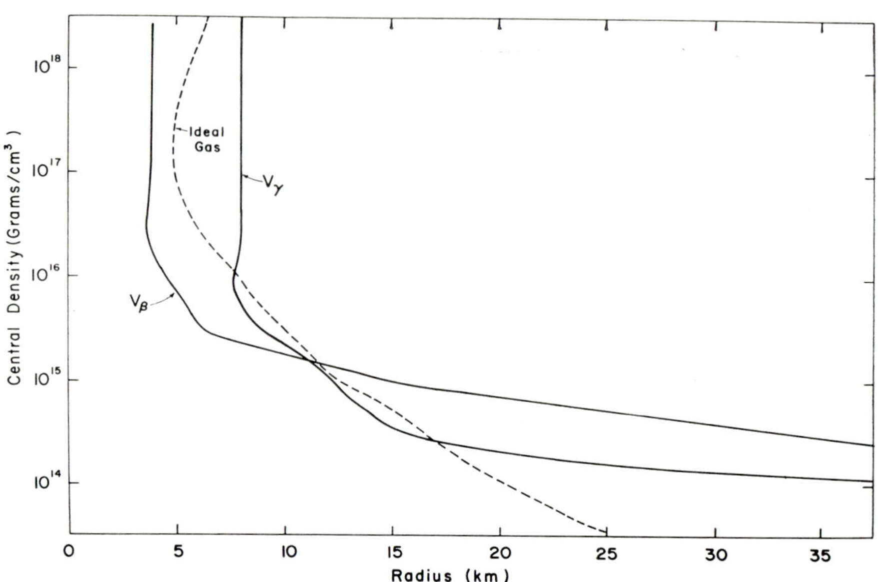

Fig. 5   Radii of the neutron star models shown in Figs. 1 and 4.

similar to that obtained assuming non-interacting particles. For V_β, the maximum stable mass peak occurs at a density of about 10^{16} gm/cm³ and a mass of about 1 solar mass. The V_γ solutions, which are probably more reliable, give a mass closer to two solar masses, with a somewhat lower density of about 2×10^{15} gm/cm³. As indicated previously, the V_γ solution is the better of the two, but the proper solution probably should lie at slightly higher density than that of V_γ, since the appearance of a new kind of baryon implies that there are new attractive forces which should soften the equation of state somewhat. It is reasonable to expect on the basis of these considerations that the stable mass limit for a neutron star will be somewhere around 1.5 solar masses. The range of radii for various models can be seen in Fig. 5; for densities of 10^{15-16} gm/cm³ the radius will be in the general vicinity of 10 km. The density variation (Fig. 6) is quite flat toward the center of the star. The relative flatness of the density distribution may have some significance with respect to the emission of gravitational radiation; if a structure with relatively little density variation from center to surface is rotating rapidly it will probably deform into a Jacobi ellipsoid, which is a structure with three unequal axes and a corresponding varying mass quadrupole moment.

If we now abandon the completely cold model, we find that the star will have an atmosphere whose properties we can compute (Fig. 7). The photospheric density will be of the order of 0.1 gm/cm^3; we note that the density will increase by many orders of magnitude within the first meter, so that an obvious feature of these neutron star models is that we are dealing with atmospheres of rather small thickness. In order to construct cooling curves, we need to know the ratio between the surface temperature and the central temperature. The temperature will be very uniform throughout the major part of the star because of the degenerate electrons, and the thermal capacity

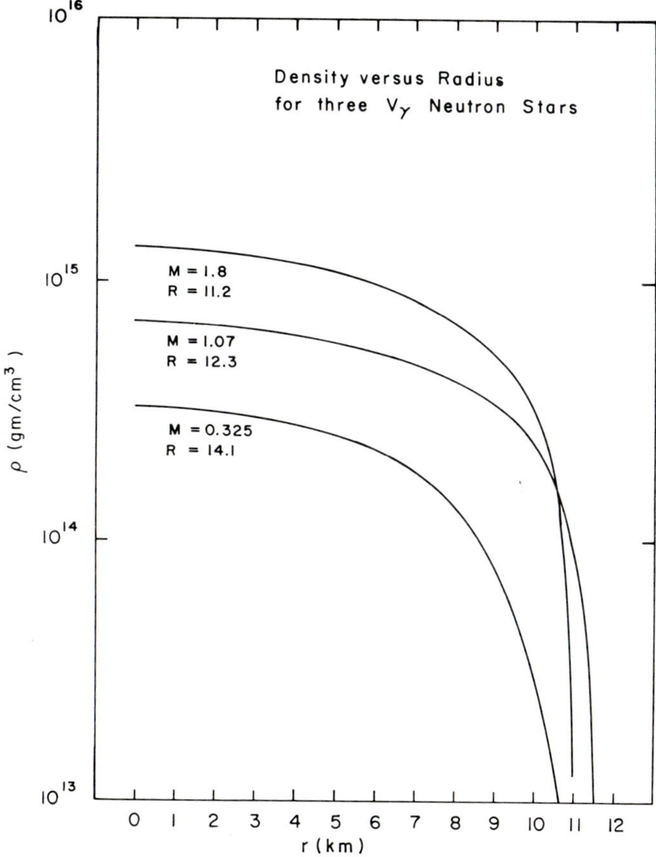

Fig. 6 Density profiles of three neutron stars based on the Levinger–
Simmons V_γ potential.

of the neutron star will be proportional to this temperature. We find that for the more interesting surface temperatures around $10^{6-7}\,°K$ the central temperature will be about 2 orders of magnitude greater than the surface temperature (see Fig. 8). At central temperatures around $10^9\,°K$ the process of neutrino-antineutrino pair emission will become very important and will be primarily responsible for the cooling of the star rather than by radiation from the surface. The cooling curves are shown in Fig. 9. Over the region in which the curves are fairly flat the cooling is due predominantly to the neutrino pair emission processes. Eventually the cooling occurs mainly by

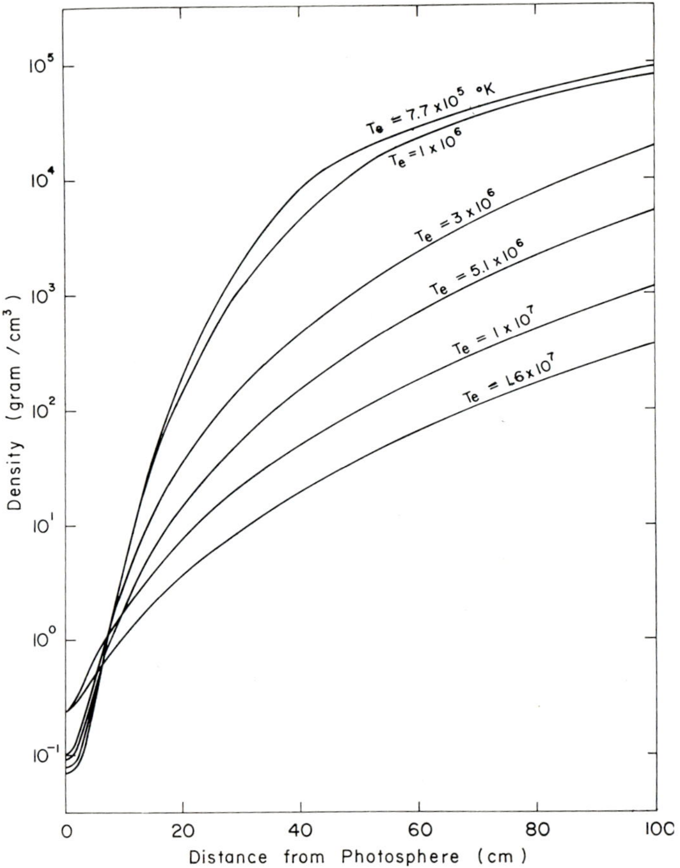

Fig. 7   Density of a neutron star atmosphere (for a radius of 10 km and a mass of one sun) below the photosphere for various surface temperatures.

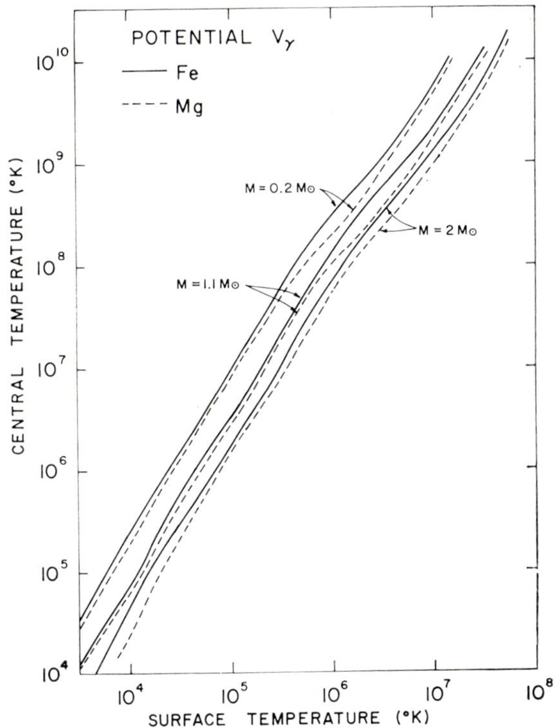

Fig. 8 Relations between surface and central temperatures of various
neutron star models.

conduction of energy to the surface and radiation from the surface, at which
point the cooling will go much more rapidly. However, if there is a magnetic
field through the surface, it will quantize the motion of the electrons, effectively
producing an oscillator potential. The states in this oscillator potential will
correspond to energy intervals of an MeV or more, which is large compared
to the thermal energies on the surface. Consequently the opacity will essentially vanish; we would then have a surface temperature effectively equal to
the central temperature and the rate of cooling would be drastically accelerated.

Finally, we turn to the question of the surface composition. Chiu and Salpeter have pointed out that the rate of diffusion into the neutron star atmosphere should be sufficiently great that hydrogen and helium will disappear
rather rapidly. Figure 10 shows the results of more precise calculations done

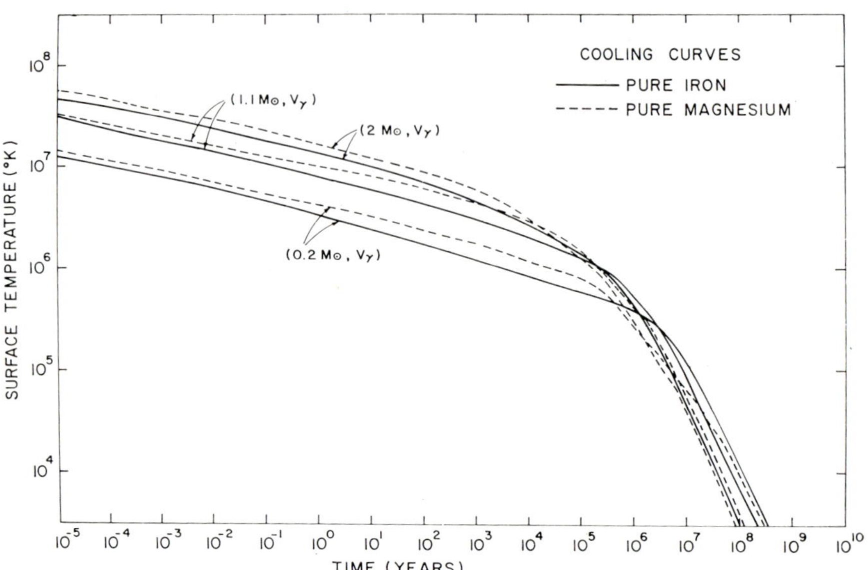

Fig. 9 Variation of surface temperature during neutrino and optical cooling of various neutron star models with either iron or magnesium surface layers.

recently by Rosen which bears out their conclusion. In his calculation, Rosen put a 150 cm layer of hydrogen on top of the neutron star, assuming that there was carbon underlying it, and determined the time for the hydrogen to disappear. We see that the hydrogen will diffuse on a time scale of the order of a few days. On this time scale, the temperature will remain near $2 \times 10^7 \,°\text{K}$, and at that temperature the hydrogen will effectively disappear from the surface layer in about a day. In the case of helium (Fig. 11) the diffusion takes a bit longer; the time scale here is more of the order of a year or so; the effective surface temperature will be closer to $10^7 \,°\text{K}$, and the helium will disappear in a few months.

Having thus summarized the relevant properties of neutron stars, we can now return to the question of energy storage. With respect to thermal energy, we see from Fig. 9 that within 10^{3-4} years the surface temperature is down to about $10^6 \,°\text{K}$, so that thermally the neutron star does not become a very strong x-ray source. The dissipation of rotational energy is dependent partly on the emission of gravitational radiation (to be discussed by Thorne) and partly on the possible effects of a magnetic field. As far as magnetic fields are concerned, if the kind of magnetic field found in ordinary stars were to be compressed down to neutron star dimensions while maintaining flux conservation, field intensities of 10^{14-16} gauss could be obtained. If the field extended radially out from the star, the rotational energy could be dissipated by expelling mass on which the field lines would exert a torque. A rough estimate can be made as to how rapidly this will slow the star; the time for the elimination of the rotation is of the order of $3 \times 10^{31}/B^2$. For fields of 10^{14-16} gauss, the rotational energy would dissipate in a matter of seconds. Thus the elimination of rotational energy is very intimately related to the question of the magnetic fields that will be present.

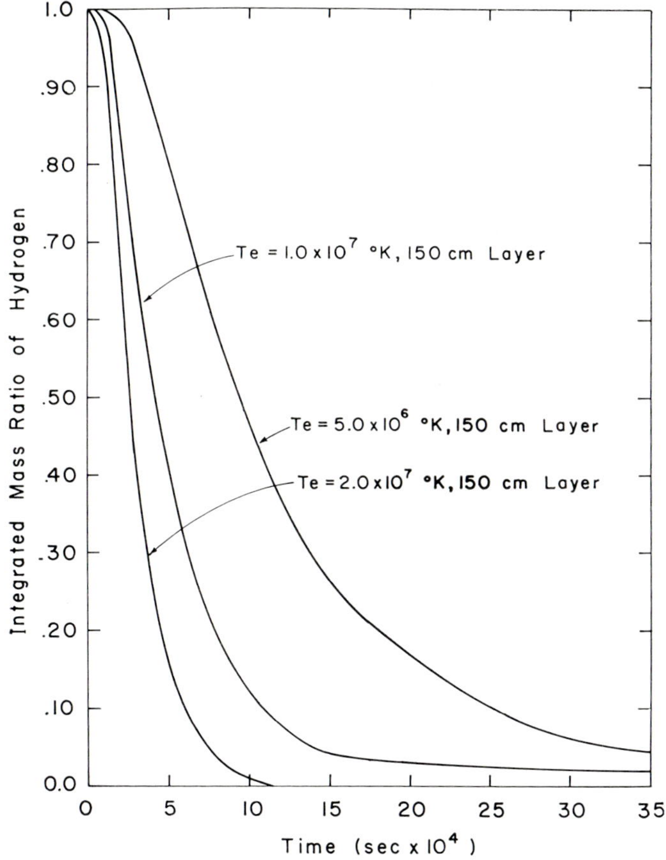

Fig. 10   Rate of depletion of a surface hydrogen layer by diffusion at various surface temperatures.

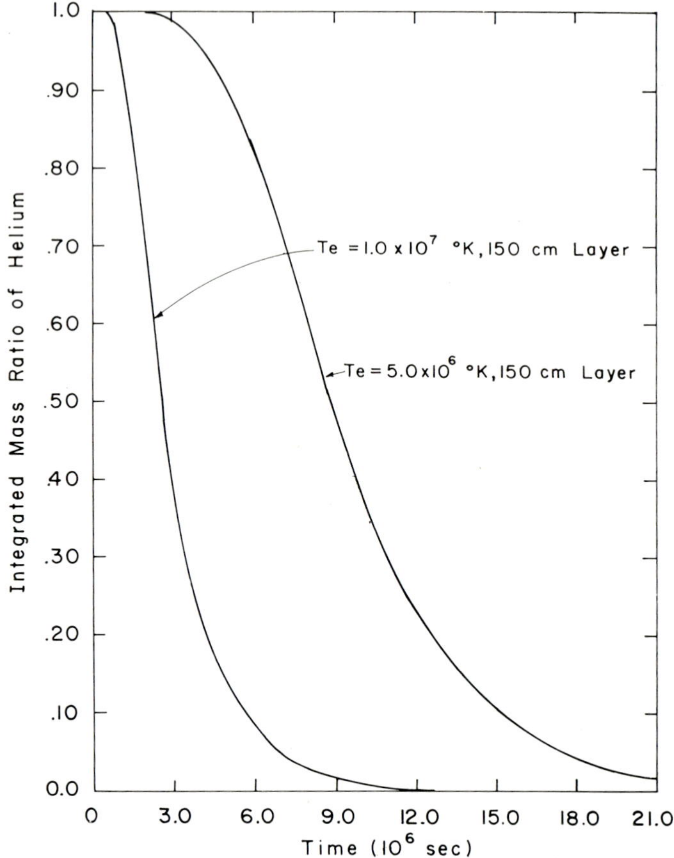

Fig. 11   Rate of depletion of a surface helium layer by diffusion at two surface temperatures.

Hoyle, Fowler, Wheeler, Finzi and I have all pointed out the possible interesting effects associated with vibrational energy storage in a neutron star. However, we have to determine how rapidly the vibration will disappear. Figure 12 shows some results of calculations done by Hansen on the vibrational energy remaining as a function of time for various neutron star models, with the energy being eliminated by the URCA process. On this basis, it seemed that out to times of the order of the age of the Crab Nebula there might be some significant amounts of vibrational energy still remaining. However, Mock looked at the energy dissipation of the vibration assuming that the vibration acts as a piston driving the neutron star atmosphere. The purpose was to study the shock dissipation of the energy in the neutron star atmosphere, leading either to enhanced heating of the atmosphere or possibly to mass ejection. The half life associated with vibrational energy dissipation by the URCA process about 100 years; it turns out that the shock dissipation of the vibrational energy has a half life of something like 30 years, so that the amount of energy that could be in vibrational form after 1000 years becomes rather uninteresting. More recently Langer has taken into account the interactions and transformations between neutrons, protons and various

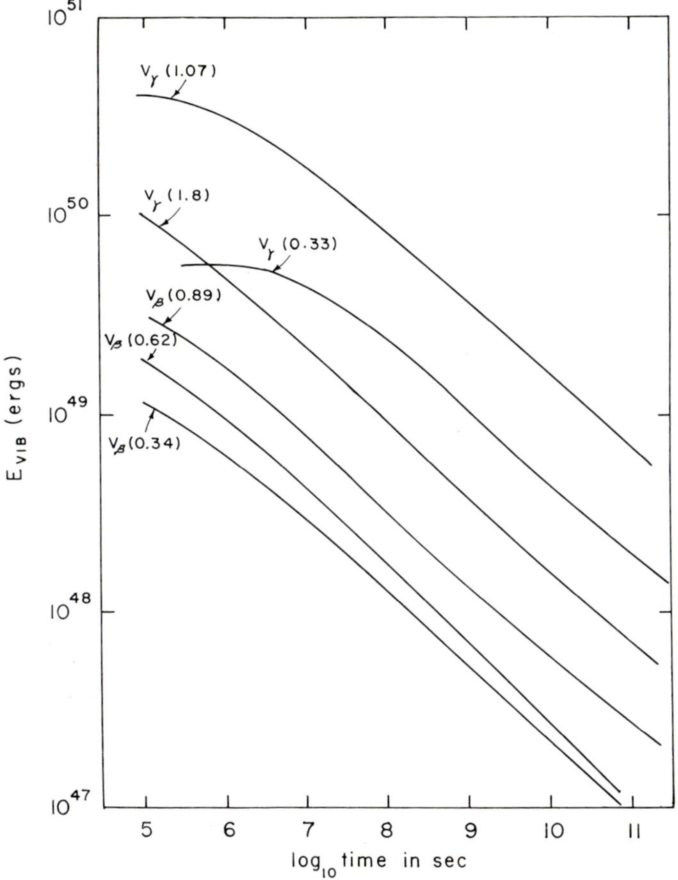

Fig. 12   Variation with time of the vibrational energy stored in various neutron star models, due to emission of neutrinos and antineutrinos.

baryons, and the conclusion reached is that if a part of the star has some baryons in it, the vibrational energy will be dumped into heat within a few seconds. If the outer part of the star which does not contain baryons is vibrating, then the waves arising from the vibration propagating into the region which has the baryons would probably result in a very rapid dissipation also.

The question of the dissipation of the magnetic energy has not yet been properly investigated. Manley has suggested that the magnetic energy may disappear on the time scale of the supernova collapse; if the magnetic energy should disappear this rapidly the thermal cooling might be slower and the rotational energy might take considerably longer to dissipate. Obviously, the magnetic properties of neutron stars need to be examined in great detail. In general, however, the conclusion seems to be that a neutron star will become a dead object rather rapidly.

Partial Bibliography of Neutron Stars

1   Neutron star models

L. Landau, 1932, On the theory of stars, *Physik. Zeits. Sowjetunion*, **1**, 285.

J. R. Oppenheimer and R. Serber, 1938, On the stability of stellar neutron cores, *Phys. Rev.* **54**, 540.

J. R. Oppenheimer and G. M. Volkoff, 1939, Massive neutron cores, *Phys. Rev.* **55**, 374.

F. Zwicky, 1938, Frequency of supernovae, *Ap. J.* **88**, 529.

– 1939, Highly collapsed stars, *Phys. Rev.* **55**, 726; also 1958, Handbuch der Physik (Berlin: Springer-Verlag) **51**, 766.

K. A. Brueckner, J. L. Gammel and J. T. Kubis, 1960, Binding energy of a neutron gas, *Phys. Rev.* **118**, 1095.

J. S. Levinger and L. M. Simmons, 1961, Neutron gas, *Phys. Rev.* **124**, 916.

V. A. Ambartsumyan and G. S. Saakyan, 1960, The degenerate superdense gas of elementary particles, *Soviet Astron.* **4**, 187.

E. E. Salpeter, 1960, Matter at high densities, *Annals of Phys.* **11**, 393.

S. Tsuruta and A. G. W. Cameron, 1965, Composition of matter in nuclear statistical equilibrium at high densities, *Can. J. Phys.* **43**, 2056.

E. E. Salpeter, 1961, Energy and pressure of a zero-temperature plasma, *Ap. J.* **134**, 669.

P. Cazzola, L. Lucaroni and C. Scarinci, 1968, Neutron stars and nuclear repulsive forces at high densities, *Nuovo Cimento* **52B**, 411.

V. L. Ginzburg and D. A. Kirzhnits, 1965, On the superfluidity of neutron stars, *Soviet Physics, JETP* **20**, 1346.

A. G. W. Cameron, 1959, Neutron star models, *Ap. J.* **130**, 884.

J. A. Wheeler, 1958, Some implications of general relativity for the structure and evolution of the universe, paper read at the Solvay Conference, Brussels, published with B. K. Harrison and M. Wakano, in La Structure et l'Evolution de l'Univers (Brussels; R. Stoops), p. 97.

T. Hamada and E. E. Salpeter, 1961, Models for zero-temperature stars, *Ap. J.* **134**, 683.

V. A. Ambartsumyan and G. S. Saakyan, 1962, On equilibrium configurations of superdense degenerate gas masses, *Soviet Astron.* **5**, 601.

V. A. Ambartsumyan and G. S. Saakyan, 1962, Internal structure of hyperon configuration of stellar masses, *Soviet Astron.* **5**, 779.

G. S. Saakyan, 1963, Comments on a paper by A. Cameron, *Soviet Astronomical AJ* **7**, 60.

S. Tsuruta and A. G. W. Cameron, 1966, Some effects of nuclear forces on neutron star models, *Can. J. Phys.* **44**, 1863.

J. N. Bahcall and R. A. Wolf, 1965, Neutron stars, *Phys. Rev. Lett.* **14**. 343.

J. N. Bahcall and R. A. Wolf, 1965, Neutron stars I. Properties at absolute zero temperature, *Phys. Rev.* **140**, B1445.

H. Y. Chiu, 1964, Supernovae, neutrinos, and neutron stars, *Annals of Phys.* **26**, 364.

2   Cooling of neutron stars

D. C. Morton, 1964, Neutron stars as x-ray sources, *Nature* **201**, 1308.

– 1964, Neutron stars as x-ray sources, *Ap. J.* **140**, 460.

H. Y. Chiu, 1965, The formation of neutron stars and their surface properties, in *Quasi-Stellar Sources and Gravitational Collapse*, ed. by I. Robinson, A. Schild, and E. L. Schucking, University of Chicago Press, Chicago.

H. Y. Chiu and E. E. Salpeter, 1964, Surface x-ray emission from neutron stars, *Phys. Rev. Lett.* **12**, 413.

A. Finzi, 1964, Neutron stars as a possible source of x-rays from outside the solar system, *Ap. J.* **139**, 1398.

S. Tsuruta and A. G. W. Cameron, 1965, Cooling of neutron stars, *Nature* **207**, 364.

– 1966, Cooling and detectability of neutron stars, *Can. J. Phys.* **44**, 1863.

A. Finzi, 1965, Cooling of a neutron star by the 'URCA' process, *Phys. Rev.* **137**, B472.

J. N. Bahcall and R. A. Wolf, 1965, Neutron stars II. Neutrino-cooling and observability, *Phys. Rev.* **140**, B1452.

– 1965, An observational test of theories of neutron-star cooling, *Ap. J.* **142**, 1254.

L. C. Rosen, 1968, Hydrogen and helium abundances in neutron-star atmospheres, *Astrophys. and Space Sci.* **1**, 372.

3   Other properties of neutron stars

C. J. Hansen, 1966, Neutrino energy losses from vibrating neutron stars, *Nature* **211**, 1069.

C. W. Misner and H. S. Zapolsky, 1964, High density behavior and dynamical stability of neutron star models, *Phys. Rev. Lett.* **12**, 635.

A. G. W. Cameron, 1965, Possible magnetospheric phenomena associated with neutron stars, *Nature* **205**, 787.

S. Tsuruta, J. P. Wright and A. G. W. Cameron, 1965, Oscillation periods of neutron stars, *Nature* **206**, 1137.

A. G. W. Cameron, 1965, Cosmic ray production by vibrating neutron stars, *Nature* **206**, 1342.

S. Tsuruta and A. G. W. Cameron, 1966, The rotation of neutron stars, *Nature* **211**, 356.

K. S. Thorne and J. R. Ipser, 1968, White-dwarf and neutron-star interpretations of pulsating radio sources, *Ap. J.* **152**, L71.

M.A.Ruderman, 1968, Crystallization and torsional oscillations of superdense stars, *Nature* **218**, 1128.

O.P.Manley, 1967, Problems relating to magnetic fields in neutron stars, *Ap. J.* **147**, 808.

J.D.Landstreet, 1967, Synchrotron radiation of neutrinos and its astrophysical significance, *Phys. Rev.* **153**, 1372.

W.-Y.Chau, 1967, Gravitational radiation from neutron stars, *Ap. J.* **147**, 664.

DISCUSSION

O. Manley   I just wanted to mention the recent reports of the disappearance of a strong x-ray source in Centaurus. It was observed in April and May 1967, during which interval it dropped by a factor of 5 or 6. It disappeared by September. Although it was not initially interpreted as being a neutron star, that interpretation, in view of your discussion, cannot be entirely excluded.

S. Colgate   Can you say anything about the gravitational binding energy of these stars?

A.G.W. Cameron   It's something like 5–10% of the mass, and not more than 20% in the most extreme cases.

G. Field   There seems to be an inconsistency in that you take into account forces between nucleons to change the equation of state, but in writing down the Fermi distribution you assume that the particles are non-interacting.

A.G.W. Cameron   This is an inconsistency that we've been very much aware of, and some beginnings have been made to correct it. I might also mention that Wolf and Bahcall suggested that another modification would be the decrease in the effective pion mass so that pions could be present in neutron stars. If so, the cooling rate would be enormously accelerated. We don't think this is correct because pions have predominantly repulsive interactions with neutrons. It appears that the effective pion mass will be increased, so they won't appear until densities of 10^{18} gm/cm^3 are reached.

F. Zwicky   What sort of gravitational red shift will be produced by a neutron star?

A.G.W. Cameron   Typically a few percent. For the maximum mass it's about 30%.

9

GRAVITATIONAL RADIATION FROM COLLAPSED SUPERNOVA REMNANTS

K. S. Thorne

California Institute of Technology
Pasadena, California

THERE SEEM to be two different kinds of supernovae under discussion at this conference: the kind astronomers see, and the kind theoreticians compute. Being a theoretician, I shall discuss the theoretician's supernovae—though we are perhaps premature in calling them supernovae until we know better their relationship to the astronomer's observations.

Let us focus attention on the gravitational collapse of massive stars which have reached the end-point of their nuclear burning. This collapse has been studied in detail by Arnett and by Colgate and White[1], among others. My approach differs from theirs in one major way: they have always idealized the collapse as being spherically symmetric; but I must include asymmetries because I am interested in gravitational waves, which will not be emitted if the star is spherically symmetric. The star may indeed be spherical initially, but the collapse phase is typically quite unstable against small non-radial perturbations[2]. Thus a star which is initially only slightly deformed will become very strongly deformed as the collapse proceeds. A time-changing quadrupole moment will develop, and gravitational radiation will be emitted. Gravitational radiation should be emitted not only during the collapse, but also after the neutron star has been formed: once the collapse is halted by nuclear forces the star will typically overshoot its equilibrium state, rebound, then pulsate non-radially with a rather large amplitude, emitting gravitational waves as it pulsates.

The gravitational radiation is important for two reasons. First, there is the possibility that we might be able to detect it here on earth; second, it represents an important damping mechanism for the neutron-star vibrations.

Let us consider the possibility of detecting the gravitational waves. The amount of energy that should be emitted as gravitational waves is of the same order as the kinetic energy in the vibrating neutron star. This is about 10^{52} ergs—about one per cent of the rest mass of the star. This energy should be emitted in a time of the order of one second. Such a burst of gravitational radiation is just the type of signal that Joseph Weber at the University of Maryland is set up to detect with gravitational radiation detectors[3]. Gravitational waves are nothing more than inhomogeneous or tidal gravitational fields which propagate at the speed of light. Consequently a gravitational wave passing through an object will cause the object to compress and distend. It is the compression and distention that Weber is attempting to measure. His most sensitive detectors operate at a frequency of about 1660 cps, corresponding to a period of about 0.6 ms. His detectors have a Q of about 10^5; i.e., for a monochromatic wave train about 10^5 waves must pass through the detector in order to excite it to its maximum amplitude. Weber has two such detectors set up, approximately a mile apart; he is measuring coincidences between them and anti-coincidences with a system of seismometers, gravimeters and tiltmeters intended to rule out any excitations due to local effects such as earthquakes, explosions, construction activity, etc. With this apparatus, he is observing approximately one event per month which meets the coincidence conditions; but, at present, he is not prepared to say that his events are due to gravitational radiation.

However, it seems rather difficult to explain what he is seeing on any other, more conventional grounds. If it is gravitational radiation which is being seen, then the only conceivable source would involve an object of about a solar mass or more, near its gravitational radius, performing some sort of non-radial motion. It could be, for example, a close binary system of two neutron stars in which the two members are collapsing into their common gravitational radius. However, it seems to be more likely that Weber is seeing supernovae of the kind theoreticians compute. We must immediately ask ourselves whether it is at all reasonable that Weber could see gravitational radiation from one "theoretician's supernova" per month, when astronomers see only one "observer's supernova" per galaxy in 300 years.

There are of the order of 10^{11} stars in the galaxy, but only about 10^8 of these stars—those which are initially about 2 solar masses or larger—are likely to collapse beyond the white-dwarf state after exhausting their nuclear fuel. Most such stars have lifetimes of the order of 10^{7-8} years. Hence, we would expect about 1–10 stellar collapses per year in the galaxy, *if* huge mass loss during the late stages of evolution does not reduce most of these stars below the Chandrasekhar limit, and *if* the massive stars are not destroyed by explosions before collapse begins. Since 10 collapses in the galaxy per year is a rough upper limit, Weber's events can all be due to theoreticians' supernovae only if his apparatus can see nearly the entire galaxy. This seems unlikely: The radiation emitted during the collapse will be spread over a very large band width with typically, perhaps, enough near 1660 cps for Weber's apparatus to detect from a distance somewhat smaller than the size of the galaxy. The radiation emitted by the pulsating neutron star will be concentrated in a narrow band width. If by chance it is at Weber's frequency, he should detect it wherever the supernova may be in our galaxy; but otherwise he cannot detect it at all.

We conclude, then, that Weber's present apparatus can probably not see a large fraction of the $\lesssim 10$ theoreticians' supernovae per year in the galaxy, but that some of his events might well be due to stars undergoing gravitational collapse.

In order to take advantage of this possibility, one wants to have better theoretical computations of the properties of the gravitational waves which would be emitted by such objects. The first such computations, so far as I know, were done by Anthony Zee and John A. Wheeler in 1965[4]. These computations were highly idealized (as have been essentially all computations to date). They treated the quadrupole pulsations of an incompressible fluid sphere of uniform density. The freqency of pulsation of the sphere was calculated using Newtonian gravitation theory; and the power emitted as gravitational waves was calculated using the linearized approximation of general relativity (i.e. the theory of weak perturbations away from flat space-time). From their model Zee and Wheeler obtained typical pulsation periods in the range from 0.2 to 10 milliseconds and damping times due to the emission of gravitational waves between about 0.01 and 10 seconds. Similar computations by Chau[5] have yielded similar results.

There are, however, some serious questions which can be raised about the idealized model used by Zee and Wheeler. First of all, realistic neutron-star models are very inhomogeneous, whereas those of Zee and Wheeler were homogeneous. Secondly, neutron stars can be remarkably compressible; the adiabatic index, Γ_1, for neutron-star models varies from about 4/3, which is by normal standards quite compressible, to 2, which is moderately incompressible. The most important difficulty, however, is the use of Newtonian theory plus the linearized (first-order) theory of general relativity to calculate the gravitational radiation. General relativity modifies the predicted Newtonian structure of neutron stars by as much as 50 per cent; and in an expansion of powers of 0.5 one cannot with confidence keep only the first-order term. As a result various people have speculated that the Zee-Wheeler results might be considerably off, and that neutron stars might even be critically damped by the gravitational radiation (damping time approximately equal to pulsation period). Other people believed that, even though relativistic effects were so strong, the idealized model ought to give fairly reasonable results. In order to settle this controversy at least three groups—Chandrasekhar; Misner and Zapolsky; and Campolattaro and myself—independently embarked, a few years ago, on much more rigorous analyses within the framework of the full theory of relativity. Since that time, Chandrasekhar, and Misner and Zapolsky have turned their attention to other problems, while Campolattaro and I, aided by discussions with the others, have continued on and have obtained some results, which I shall now summarize[6].

We begin by taking a realistic, fully relativistic neutron-star model. We perturb the star, and describe the perturbation by the vector displacement of the fluid from its equilibrium position. We describe the gravitational field in terms of a perturbation in the general-relativistic metric tensor. We consider the problem only within the framework of first-order perturbations away from the fully-relativistic equilibrium configuration. We divide the perturbations into spherical harmonics. According to group-theoretic arguments there will be no mixing between the various harmonics. Hence, we can deal independently with each of the various harmonics characterized by the usual indices l and m and the parity π. From the perturbed stress-energy associated with these perturbations, we calculate the perturbed Einstein equations which govern the motion of the pulsating star.

The perturbed Einstein equations are a system of coupled hyperbolic differential equations for the fluid motions and gravitational waves. Rather than integrating these equations numerically, we divide the general pulsation of the star into periodic, standing-wave normal modes. These normal modes correspond to the following idealized experiment. We imagine that we send into the star, from radial infinity, gravitational waves periodic with some precise frequency. Some of the gravitational-wave energy excites the star into pulsation, while the rest bounces back toward infinity. Near radial infinity we set up a perfect reflector of gravitational waves, which reflects the outgoing waves back in. Eventually we obtain a system of standing waves between the star and infinity, coupled to a system of standing pulsations inside the star. These are our standing-wave normal modes.

Recently I have calculated the details of such standing-wave normal modes for several realistic neutron-star models. For each stellar model, and for each choice of spherical harmonic, it is useful to construct a graph similar to Fig. 1. Such a graph plots, as a function of angular frequency, the energy of pulsation of the matter in the star divided by the energy in one wavelength of the waves outside the star. Very sharp resonances appear at certain frequencies. What is happening at these resonances is that almost all of the gravitational-wave energy goes into the star because the frequency is that of a quasi-normal mode of pulsation of the star. ("Quasi-" because radiation damping makes the motion not quite periodic.)

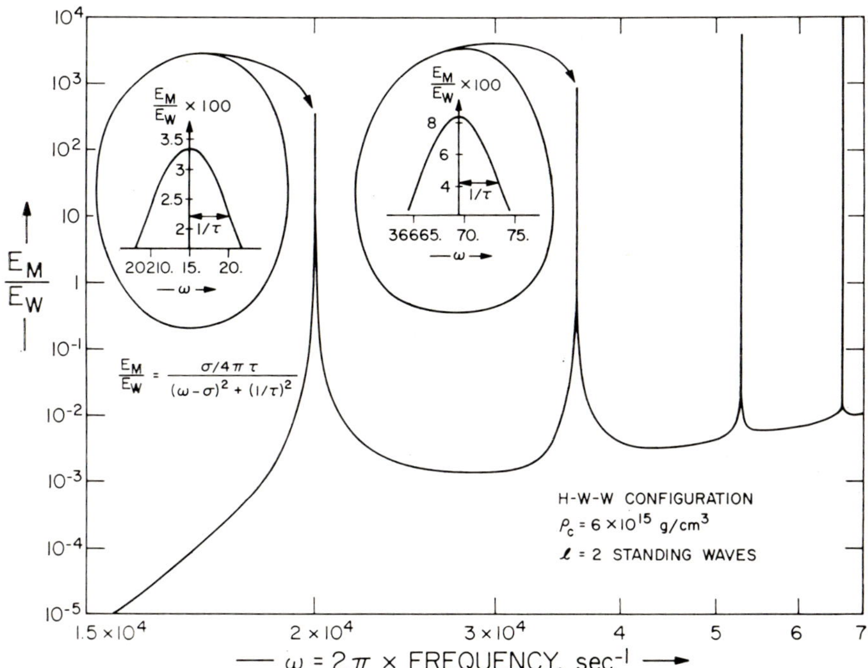

Fig. 1 Resonances in the quadrupole ($l = 2$) standing-wave normal modes for a H–W–W neutron-star model of central density 6×10^{15} g/cm^3. Plotted as a function of ω is the ratio of the total pulsation energy, E_M, of the matter in the star, to the energy, E_W, in one wavelength of the standing gravitational waves.

By wave-packet techniques familiar from quantum theory one proves the following: If the star is set into pulsation in one of its quasi-normal modes, it will subsequently emit purely outgoing gravitational waves that will damp its motions exponentially in time,

$$\text{Amplitude} \propto e^{-t/\tau}.$$

The damping rate, $1/\tau$, for these realistic pulsations is precisely equal to the half-width of the resonance in a standing-wave plot such as Fig. 1. Consequently, by studying the standing waves we can obtain the normal frequencies of pulsation of the star and the half-life for the damping of these pulsations by the emission of gravitational radiation. The law of energy conservation will then tell us the power radiated.

I have performed this analysis for a number of models. The results are presented in Table 1. Table 1 indicates that the periods of pulsation and the radiation damping rates are of the same order of magnitude when calculated by the sophisticated method as when calculated from linearized general relativity. Note that the damping is more rapid for the lower harmonics than for the higher harmonics; also, the higher the central density the more important general relativity is, and the more rapid the damping.

From Table 1 we see that any non-radial pulsations in a neutron star should damp out in a time of the order of a second. The radial pulsations will also damp out rapidly for the following reason: The star is very likely to be rotating and will therefore be deformed from spherical symmetry. This deformation will couple the quadrupole modes to the radial modes. One can calculate that for reasonable periods of rotation the radial modes will be damped in a time shorter than or of the order of a few hours[7].

With respect to Weber's experiment and the likelihood that he will see a pulsating neutron star, we can say that he is working at a frequency which is approximately right, although perhaps a bit too low for the more massive stars. His Q (i.e., the ratio of damping time to period) is somewhat high for detection of radiation from neutron-star pulsations. A Q of the order of 1000 would be more reasonable than 10^5. We should remember, however, that Weber can also observe gravitational waves from the star as it collapses, and that he will probably have a better chance of seeing these gravitational waves than those from the pulsating star.

Finally, with respect to hopes voiced in the past[8] that a supernova remnant might store large amounts of energy in neutron-star pulsations for a long

Table 1 Quadrupole pulsations for three neutron-star models

Equation of state	Harrison–Wheeler[a]				V_γ[b]		V_γ[b]		
Central density (g/cm^3)	6×10^{15}				5.15×10^{14}		3×10^{15}		
Mass (solar masses)	0.682				0.677		1.95		
$2\,GM/Rc^2$	0.240				0.157		0.58		
Pulsation mode	0	1	2	3	0	1	0	1	2
Period T (msec)	0.310	0.171	0.119	0.0938	0.699	0.236	0.378	0.156	0.103
Damping time τ (sec)	0.19	0.29	–	$\gtrsim 5$	1.4	$\gtrsim 10$	0.17	1.8	2.6
τ/T	580	1700	–	$\gtrsim 50{,}000$	2000	$\gtrsim 40{,}000$	450	12,000	26,000

[a] The Harrison–Wheeler equation of state leads to Harrison–Wakano–Wheeler (H–W–W) models. (See Harrison, Thorne, Wakano, Wheeler, *Gravitation Theory and Gravitational Collapse*, U. Chicago Press, 1965.) This H–W–W neutron star is the one of maximum mass and maximum central density. See Fig. 1 for the data from which these results were deduced.

[b] The Levinger–Simmons V_γ equation of state leads to Tsuruta–Cameron V_γ models. (See S. Tsuruta and A. G. W. Cameron, *Can. J. Phys.*, **44**, 1895.) Our second V_γ neutron star is the one of maximum mass and maximum central density.

period of time, let me re-emphasize this: These computations indicate that a neutron star should be pulsationally very dead within a few hours after it forms.

Acknowledgments

I thank Barbara A. Zimmerman for assistance with the numerical calculations; I thank A. G. W. Cameron, A. Campolattaro, S. Chandrasekhar, C. W. Misner, and J. A. Wheeler for numerous helpful discussions; and I thank P. J. Brancazio for assistance with the preparation of this written version of my paper.

References

1. See their contributions to these proceedings.
2. See, e.g., C. Hunter, *Astrophys. J.*, **136**, 594 (1962) and **139**, 570 (1964); also C. C. Lin, L. Mastel, and F. H. Shu, *Astrophys. J.*, **142**, 1431 (1965); also M. Fujimoto, *Astrophys. J.*, **152**, 523 (1968).
3. For summaries of Weber's experiment see J. Weber, *Phys. Rev. Letters*, **20**, 1307 (1968), and references cited therein.
4. For a summary of results see J. A. Wheeler, in *Ann. Rev. Astron. Astrophys.*, **4**, 393 (1966).
5. W.-Y. Chau, *Astrophys. J.*, **147**, 664 (1967).
6. For full details of our analysis see a series of three papers in the *Astrophysical Journal*: I—K. S. Thorne and A. Campolattaro, *Astrophys. J.*, **149**, 591 (1967); II—R. Price and K. S. Thorne, *Astrophys. J.*, in press; III—K. S. Thorne, paper in preparation.
7. For rough, linearized-theory calculations see J. A. Wheeler, Ref. 5. For sophisticated calculations with fully relativistic stellar models, see papers III and IV in a series by J. B. Hartle, S. M. Chitre, and K. S. Thorne, to be published in the Astrophysical Journal.
8. F. Hoyle, J. V. Narlikar, J. A. Wheeler, *Nature*, **203**, 914 (1964); A. G. W. Cameron, *Nature*, **205**, 787 (1965); A. Finzi, *Phys. Rev. Letters*, **15**, 599 (1965).

DISCUSSION

Voice You stated that your calculations included only first-order perturbations. How does this differ from the linearized theory?

K.S.Thorne In the linearized theory one perturbs away from a Newtonian picture. Here we are perturbing away from a fully relativistic star which is fairly near its gravitational radius. We believe that our results are reasonable to the order of $(\xi/R)^2$, where ξ is the amplitude of the pulsation and R is the radius of the star, independently of how compact the star may be.

H.Reeves How much energy is emitted?

K.S.Thorne This will depend on the amplitude of pulsation. For $(\xi/R) = 0.3$, the energy emitted by a typical neutron-star model would be of the order of 10^{52} ergs.

Voice Given the very small relative line width of the gravitational radiation that your models predict and the narrow instrumental response of Weber's detector, it is not clear to me why there is any likelihood of detection.

K.S.Thorne Your remark is basically correct; Weber is much more likely to detect the radiation from the collapsing star because that radiation will have a wide-frequency spectrum. On the other hand, the frequency of the neutron star pulsation will not be that sharp. It will in fact be a weak function of the amplitude, so that as the pulsation is damped, a moderately wide band of frequencies will be produced.

C.W.Misner It turns out that if the pulse sent out by the star has a fairly wide band-width, then the total energy entering Weber's apparatus will not depend on the Q of the apparatus. The reason is that the cross section for observing the energy is proportional to Q while the frequency range within the band-width of the apparatus decreases as $1/Q$. Therefore the total energy input to the apparatus is independent of Q. However, the signal-to-noise ratio will improve with increased Q. Unless the wave shape can be known in advance, then, it is advantageous to have a high Q. It is critical, however, that the energy come in as a pulse which is short compared to the damping time of the apparatus.

R. Macklin   Is it reasonable to expect that Weber-type detectors could be developed to a state where one could find the phase lag and obtain a direction?

C. W. Misner   The maximum time resolution that Weber expects to be able to achieve by reasonable modifications of the present apparatus is about 1 msec, so phase resolution would not be feasible.

10

FLUORESCENCE THEORY
OF SUPERNOVA LIGHT

L. Sartori and P. Morrison

Massachusetts Institute of Technology
Cambridge, Mass.

THE "THEORETICIAN'S SUPERNOVA", which has been much discussed at this conference, is unfortunately not subject to direct observation. Our concern is not with this esoteric object, but rather with the "observer's supernova", characterized by light curves and spectra. The model we shall describe accounts satisfactorily for these measured properties, particularly in the case of Type I supernovae. Our considerations imply that the theoretician's and observer's supernovae are only weakly coupled, the behavior of the latter being determined primarily by the medium that surrounds the exploding object. If these considerations are correct the observations do not constitute a sensitive test for a theory of the explosive event, but provide only a set of boundary conditions in the form of requirements as to energy release, mass ejection, and the like.

We assume that the explosion proper takes place over a very short time—a day at most—and that a substantial fraction of the primary energy release is in the form of ultraviolet continuum in the energy range between ~ 25 and 100 eV. We can then account for the observed light as fluorescence excited by this radiation impinging on the material that surrounds the explosion site. The central point in the argument is that the light seen by an observer at a given time was not all emitted at the same time, but is instead a composite of radiation emitted at various times from those points Q for which the total path length from source to Q to observer is a constant. The locus of points that satisfy this condition is an ellipsoidal shell, with the supernova and observer as foci, that expands with time. The effective luminosity L inferred by the observer from his measurements at a given time is an integral over the ellipsoidal source region; the time dependence of L therefore reflects the attenuation of the exciting pulse as it travels out through the medium.

Consider first a single fluorescent line excited by the primary flux. Its contribution to $L(t)$ depends upon the concentration of the emitting atom or ion, and on the cross section for the absorption that excites the fluorescent line in question. In the simplest case of uniform density and an absorption line of square shape, one obtains after a straightforward derivation the result:

$$L(t) = KE_1 \left(\frac{ct}{2\Lambda} \right) \tag{1}$$

where t is measured from the observer's first notice of the explosion. Here K is a constant; $E_1(x)$ is the exponential integral, a well-studied function that behaves asymptotically as e^{-x}/x and for small x is approximately $\log(1/x)$; Λ is the mean free path for the exciting photons, and is given by:

$$\Lambda = \left(\frac{mc}{\pi e^2} \right) \frac{\Delta v}{nf} \tag{2}$$

where Δv is the effective width of the absorption line, n is the concentration of absorbers, and f the oscillator strength for the absorption.

Since the medium is in primarily radial motion (at the very least it is being pushed out by the radiation pressure of the exciting pulse), the emitted fluorescent photons are Doppler-shifted. The spectral line is therefore broadened; it is readily shown that the line is also, in general, shifted. For the shape of the line depends upon the relative amounts of radiation received from the various parts of the source ellipsoid; the photons emanating from the near side of the source are blue-shifted, whereas those emanating from the far side are red-shifted. But the strength of the fluorescent emission decreases with distance from the primary source. Most of the radiation detected at observer time t emanates from points whose distance from the explosion site is between $ct/2$ and $ct/2 + \Lambda$. In the early stages of observation, when $ct/2 \ll \Lambda$, most of the observed emission comes from the near side of the source, and consequently the maximum of the line should be blue shifted. Later, more of the effective emission region is on the far side and the line should be observed as red shifted. The theory thus predicts that every line should red shift with time; this red shift is, at least at first, a diminishing blue shift. The rate at which the shift occurs depends on the value of Λ. If Λ is small, the change from the regime $ct/2 \ll \Lambda$ to the opposite regime $ct/2$

$\ll \Lambda$ takes place over a short period of time; if Λ is large, the change may be so slow as to be undetectable over the available period of observation.

If He II is a major constituent of the fluorescent medium for Type I supernovae, then other visible lines of this ion are also emitted. These lines constitute the Brackett series $n \to 4$ ($n = 6, 7, \ldots$). We can estimate the theoretical relative intensities of the lines; but whether or not a given line is actually discernible depends also on the possible presence of overlapping lines. Our study of the spectra convinces us that they are indeed attributable largely to He II, particularly in the photographic region. Each Brackett line up to about $n = 10$ can be discerned, with varying degrees of confidence, in at least some of the spectra we have examined, and all but one of the major photographic features is thus accounted for, at least in part. Not every line can be seen in every spectrum, and we have not checked the relative intensities very carefully. The spectra are very complicated; undoubtedly some of the bands are composite, as Professor Minkowski has emphasized, with contributions from other species; and there are peculiarities that appear only in particular cases. We have not concerned ourselves with fine details, feeling that essential gross features of the spectrum must first be understood. It is our conviction that this has now been accomplished.

The predominance of the He II lines in the photographic region helps explain why the light curve exhibits such a simple behavior; that is, why it is fitted by a single function of the form (1). Notice that all the levels of He II with $n > 4$ lie within some $3\frac{1}{2}$ volts of the ionization energy. With the particle velocities indicated by the emission widths, Doppler broadening makes the absorption lines to these levels overlap strongly. Therefore in calculating the attenuation of the photon flux all these absorption lines must be treated as a single band, characterized by a single value of Λ determined from the over-all effective width and the total f-number. It follows that the Brackett emission lines excited by these absorptions all decline with the same time dependence. In addition, every Brackett emission leaves the electron in the $4s$ or the $4d$ sublevel, from either of which it has a good chance of cascading down via $n = 3$, thus emitting a Paschen α photon. (Direct transition to the ground state, normally the most probable, is in this case forbidden.) As a result of the cascade mechanism, the Paschen α emission acquires a substantial additional intensity which is directly synchronized to the emission in the Brackett lines.

The non-helium part of the spectrum is presumably due to lines of C, N, O, Ne, and possibly even heavier ions. (Any emission from hydrogen would only enhance the even members of the He II Brackett series.) As we have already remarked, the red part of the spectrum contains more of these non-helium features than does the blue. The visual light curve, which is sensitive to this spectral region, differs from the photographic but not overwhelmingly so. (According to Mihalas, the best exponential fit to the tail of the visual light curve of IC 4182 has a lifetime about 50% longer than the corresponding quantity computed from the photographic light curve.) Within the context of our theory, this means that the pertinent values of Λ for the lines involved are not too different from the value for He II. The concentrations of the ions C, N, O etc., are surely lower than that of helium; but the values of f and $\Delta\nu$ in particular instances can easily be such as to make the Λ determined by Eq. (2) comparable to 50 or 100 days. There are plausible candidates for the major red features, but we propose no definite identifications.

Substituting into Eq. (2) the value $\Lambda/c \approx 50$ days obtained from the fit to the light curve, and using the appropriate f-numbers for He II, we obtain a value $n \approx 4\,\mathrm{cm}^{-3}$ for the concentration of helium in the fluorescent medium in the range $10^{17}\,\mathrm{cm} \gtrsim r \gtrsim 10^{18}\,\mathrm{cm}$. This concentration is clearly greater than one can expect to find in the normal interstellar medium. We suggest therefore that the medium has in fact been provided by the exploding star itself through mass ejection during the pre-supernova stages. The total amount of material needed to provide the required concentration in a sphere one light year in radius is about $0.1\,\mathrm{M}_\odot$, which is quite plausibly ascribed to mass loss. (It is comparable to the masses of planetary nebulae.) Moreover, the hypothesis that the fluorescent medium has been provided by the pre-supernova makes it easier to understand the striking uniformity of Type I supernovae. If one had to depend on ambient gas for the fluorescent medium, one would be hard put to explain why these objects are so similar despite the fact that they occur at such widely different locations in many types of galaxies.

The energy requirements implied by our model are readily estimated. Integrating under the photographic light curve we arrive at a total energy $\sim 3 \times 10^{48}$ erg; this represents observed light only, with no bolometric correction. To excite visible fluorescence in He II requires a photon of about 50 eV. But an optical transition will not occur each time the appropriate initial state is excited; the efficiency (fluorescent yield) is in fact only about 5%: some 20 ultraviolet photons are required to produce one visible fluorescent photon. With this efficiency factor taken into account, the observed fluorescence implies $\gtrsim 10^{51}$ erg in primary ultraviolet energy. Moreover, the latter figure represents only the energy in the band between 50 and 55 eV. We have no idea as to the shape of the primary spectrum; but it would surely be unreasonable to suppose that all the primary flux is concentrated in so narrow a spectral region; as a rough estimate we multiply by another order of magnitude to allow for the remainder of the spectrum. Thus we arrive at an estimate in the range 10^{52} erg for the energy which must be provided as ultraviolet continuum by the primary explosion. Only a small part of this energy is finally emitted as visible fluorescence; most of the remainder escapes and presumably creates a huge H II region. Such HII regions would last a long time after the energy source is turned off, and ought therefore to be associated with many supernova remnants; the emission measures could be such as to make the H II region observable, and indeed some recent observations may be subject to such an interpretation.

A critical problem in the model is the question of the ionization of the helium. From our estimates of the density and of the amount of fluorescence emitted, we conclude that the average helium ion must emit $\sim 10^5$ fluorescent photons; it must do this without being ionized a second time, since the recombination times are very long. If the ultraviolet spectrum that impinges on the fluorescent region were continuous across the second ionization edge at 54 volts, ionization would occur after at most a few excitations. One must conclude, therefore, that the ionizing flux is not present in the fluorescent medium, or at least is delayed until the fluorescent emission has been completed.

There exists a mechanism that plausibly provides the desired filtering action. Suppose there is a region, fairly close to the explosion site, in which the density is high enough to make the recombination time short (say $n \gtrsim 10^{12}$). The intense ionizing flux impinging on such a region produces an analogue of the familiar Stromgren sphere: ionization practically complete out to some distance r_0, then falling abruptly almost to zero across a narrow boundary layer. The ionizing radiation is absorbed because of the finite probability ($\sim \frac{1}{3}$) of recombination to an excited state of He^+ rather then directly to the ground state. Each time this happens, an ionizing photon is effectively converted into two or more photons below the ionization edge. Provided the dense region is thick enough, practically no ionizing radiation escapes. This is precisely the mechanism by which the Lyman continuum is filtered in H II regions and planetary nebulae. The density requirement can be expressed in the form

$$n^2 r_0^3 = F$$

where F depends on the intensity of the ionizing flux and is of course much higher for the supernova than for H II regions or planetary nebulae. It follows that the total mass of the filtering region scales as $r_0^{3/2}$. If we impose an upper limit ~ 1 $M_\odot$ on the total mass, then from our estimate of the ionizing flux r_0 must be $\lesssim 10^{14}$ cm. This is not an unreasonable requirement.

The presence of a dense layer also provides a plausible explanation for the early part of the light curve. The characteristic time of the initial rise and fall, ~ 10 days, seems too long to be a direct manifestation of the primary explosion. (The rise is markedly slower than that of some fast novae; this is particularly manifest when the luminosity is plotted on a linear rather than a logarithmic scale.) Moreover, the appearance of the early spectra suggests that at least a substantial part of the radiation has a common physical origin with that observed later on; the characteristic 4686 band is sometimes discernible even before maximum. We therefore consider the possibility that even the very early light represents primarily fluorescent emission.

The simple theory that leads to a light curve of the form $E_1(x)$ is based on a δ-function primary input, but may be generalized to an input function of arbitrary time dependence. It is readily shown that if the input lasts a finite time T, then the light curve is monotonically decreasing for all $t > T$, regardless of the density distribution. Hence, an adequate explanation of the early light curve requires an input function whose duration is ~ 10 days. We propose that the ultraviolet pulse, whose characteristic time is initially much shorter, is spread out in time as it passes through the dense ionized region described above. The mechanism principally responsible for the spreading is Thomson scattering. With the density we envisage for the dense layer, the layer is many mean free paths thick for Thomson scattering. Thus the photons diffuse through it, and the input pulse that excites the fluorescence further out is effectively broadened. With a diffusion-type input, the light curve predicted by the theory agrees qualitatively with the observations. The analysis is complicated because the early light curve is sensitive also to density variations in the inner part of the fluorescent medium ($r \sim 10^{17}$ cm). Investigation of this aspect of the problem is continuing.

We have presented what we feel is a consistent model for Type I supernovae, and are now trying to extend the model to those of Type II. Unfortunately, the data on the latter are limited: the light curves are observed for shorter periods of time, and there are few available spectra. Furthermore, Type II supernovae form not nearly so homogeneous a class as do Type I. Nevertheless, it seems reasonable to us that the same fluorescent mechanism operates in both types. The spectra have the same general appearance, and the Type II light curves, although complicated, share one important property with those of Type I: they decrease monotonically after the initial rise, i.e., they show no secondary maxima, as the light curves of novae frequently do. This property is a natural consequence of the model, provided the absorbing material is not distributed in too pathological a manner.

The uniformity of Type I supernovae and their occurrence in all types of galaxies have been explained by the proposition that these objects create their own fluorescent medium. It is then tempting to suppose that supernovae of Type II eject much less material and must therefore make do with the ambient medium. This hypothesis explains naturally why Type II are seen only in those galaxies that are rich in gas, viz. spirals. Also, since the normal interstellar medium is largely hydrogen, one would expect to see fluorescence from H I appearing strongly. And indeed the Balmer lines are prominent in the spectra of Type II supernovae. The differences among the light curves of SN II according to this picture reflect differences in local environment at the places where the explosions occur; and the fairly abrupt changes of slope of an individual light curve can be understood in terms of the source ellipsoid encountering an interstellar gas cloud. The useful part of the primary flux is now in the range $\sim 10\text{--}15$ eV., which means that each optical photon now carries off about a quarter of the energy of the exciting ultraviolet photon. Moreover, the fluorescent yield for the Balmer series is about 12%; the over-all efficiency for converting primary ultraviolet flux into visible fluorescence is thus about an order of magnitude higher in hydrogen than in He II. Since the apparent visible energy in Type II supernovae is comparable to (perhaps somewhat smaller than) that of Type I, the lower bound set by our model on the energy requirement is at least an order of magnitude smaller for the Type II.

There are, of course, many more details of the Type II phenomenon to be understood. At the moment, our need is for more Type II spectra.

DISCUSSION

A. G. W. Cameron　One of the interesting things in this model is the need for a shell of higher density material surrounding the star which will delay the ionizing photons. Dr. Colgate has discussed in his models the mass which comes out relativistically and the associated gamma pulse. I wonder whether the collision between the relativistic mass and the high density medium, and the absorption of the gamma rays, could provide the primary energy source. How much energy is there in the relativistic mass fraction and the gamma pulse?

S. Colgate　There are about 10^{49} ergs in the relativistic fraction in the more optimistic models. However, if we include the energy from the relativistic fraction down to 10 MeV/nucleon, the total might be as high as 10^{52} ergs.

R. Minkowski　I would like to make some comments about the spectroscopy. First of all, the suggestion that the strong band in the blue is 4686 Å is a reasonable one, although during the earlier phases it may be that there are additional lines involved. However, I do not believe that it is in any way necessary to interpret the spectrum as indicating the presence of any other He II lines. The interpretation that the whole spectrum is mainly He II meets difficulties. In the blue part of the spectrum all features shift towards the red with increasing time, but in the red part of the spectrum there are features which are present right from the beginning and never shift in wavelength. This suggests different origins for the features in the blue and red parts of the spectrum. There are transformations in the spectrum; some lines fade out, while others fade out and then come back later on. The most pronounced changes seem to occur after about 3 or 4 weeks, when the light curve changes from the initial decay to a quasi-exponential fall-off. After 4 weeks we see a remarkable phenomenon: the whole blue part of the spectrum fades out, only to brighten up again a week later. In the very late phases, the red part of the spectrum will fade. Of course, if one examines a spectrum at a given phase one may be able to see features apparently corresponding to some of the He II lines, but the time dependence of these features presents a considerably more complicated picture.

I would like to illustrate the situation further by showing two spectra, one of a Type I supernova and one of a Type II (Fig. 1). For the Type II we see

the strong Hα line at 6563 Å, accompanied by Hβ (4861 Å) and Hγ (4340 Å).
For the Type I we see the persistent red band, which is certainly not due to
He II. The band at 4686 Å (the 4 → 3 transition in He II) can be seen in both
spectra. However, the Type II spectrum is quite weak around 5412 Å (6 → 4
transition in He II). Similarly, in the Type I spectrum there is a conspicuous
minimum at 4861 Å (8 → 4) transition in He II). There is a line near 4340 Å

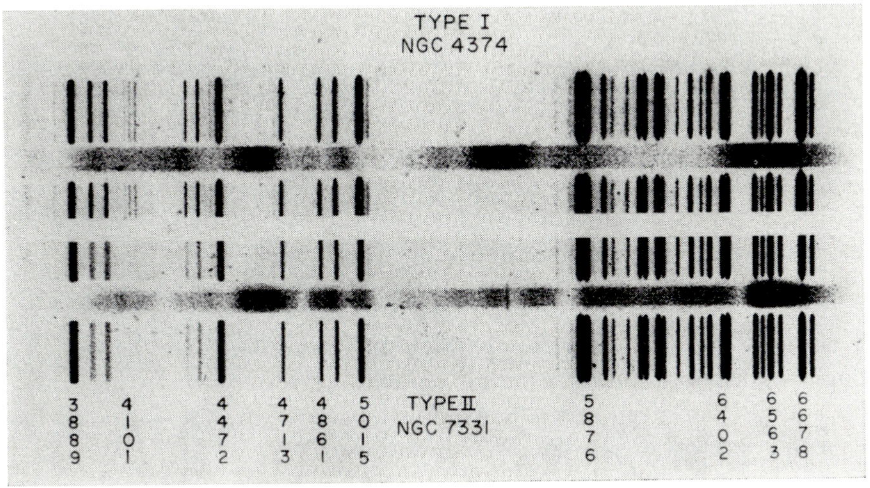

Fig. 1   Spectra of Supernovae NGC 4374 (1957), Type I, about 25 days
after maximum, (above) and NGC 7331 (1959), Type II, 60 days after
maximum.

10 → 4) but it is shifted in the wrong direction. There is nothing recognizable)
around 4200 Å (11 → 4), and nothing conclusive can be said about the
general intensity near 4101 (12 → 4). It seems clear that except for 4686 Å,
He II is not present, and I would say that it corresponds to general spectro-
scopic experience that 4686 Å is usually a very strong line, while other lines
associated with He II are much weaker. Furthermore, I do not think that it is
important that these lines be observed in order to explain the light curve.
As I understand it, all that matters is that all lines have conditions of excitation which are very similar to those for the 4686 Å line.

L. Sartori   We agree with Dr. Minkowski that these lines are going to be
weak. The identification of these weak lines is not essential; however, if they
do appear it is consistent with their being the He II lines. We do not claim

to have accounted for all the lines that are seen, but all of the spectra we have examined are consistent with the idea that these lines are present along with any other lines that might be produced.

F. Zwicky   It seems to me to be completely wrong to think of the band around 4686 Å as the result of a single spectral line. It is considerably more complicated. We have found patterns of very rapidly changing superimposed peaks on this band which seem to indicate that several lines may be involved.

S. Colgate   Is it possible to explain the excitation of the medium in terms of the relativistic particles that are emitted in the explosion?

P. Morrison   We found that if we used relativistic particles as the source of the excitation, we could not reproduce the light curve. Furthermore, if we take into account the number of collisions that a relativistic particle will have in passing through the medium, we find that the particle will give up only a small part of its energy.

G. Field   In accounting for the high-density medium around the star, don't you run into difficulty in that any reasonable outflow of material will have a $1/r^2$ dependence, so that the helium concentration would in fact depend very critically on the distance to the star?

L. Sartori   The distribution of matter will depend on many factors, such as the rate of ejection, velocities, and so on. Furthermore, the problem of inverting the density-luminosity relation does not have a unique solution. That is to say, even if the light curve had exactly an $E_1(x)$ dependence, it would not require uniform density. We have shown that the density distribution

$$n(r) = \frac{n_0}{1 + Ce^{r/A}}$$

with C arbitrary, gives identically the same light curve as does a uniform n_0. Thus it is clear that a considerable departure from uniformity can be tolerated without spoiling the agreement with observations. (But not, of course, a $1/r^2$ density distribution.) We are studying this question further.

P. Morrison   The solar wind certainly doesn't fall off as $1/r^2$. There does not seem to be any smooth dependence, and it may even pile up beyond the orbit of Jupiter.

A.G.W.Cameron I think the general picture that would be required is one in which well before the explosion there is rapid mass ejection which decreases with time and then increases again just before the explosion to give the moderating shell. Also, if this shell absorbs the energy of the relativistic particles and initial gamma pulse, the prompt energy released might be used to explain the maximum in the light curve.

P.Morrison That possibility seems reasonable. However, our basic approach has been to work backwards from the end of the direct observations — about 3 years — to about 5 or 10 days after the explosion. There's still a wide gap between us and those people who start with neutron cores and work outwards.

A.G.W.Cameron We should keep in mind, in view of the previous discussion on gravitational wave detection, that if we see only those supernovae which have created the right kind of environment previous to the explosion we may be forced to re-examine our estimates of the frequency of explosions in galaxies.

11

X-RAY EMISSION
FROM SUPERNOVA REMNANTS

P. Morrison

*Massachusetts Institute of Technology
Cambridge, Mass.*

THE RECENT ADVANCES in x-ray astronomy have led us to re-examine supernova remnants from a different point of view; namely, we are now considering them as x-ray sources. In this paper we would like to propose a model in
which the x-ray emission of a supernova remnant can be connected in a
plausible manner to the supernova explosion itself.

Let us begin by examining the Crab Nebula in detail. Figure 1 shows the
spectrum; it follows a power law in the radio region and then bends sharply

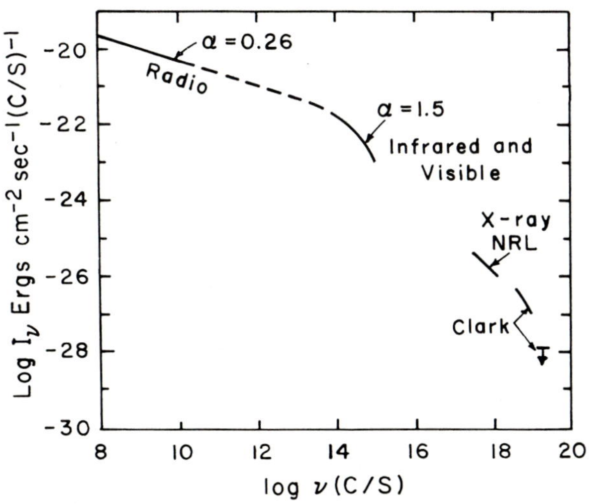

Fig. 1 The spectrum of the Crab Nebula.

in the visible and infrared. The x-ray data can, if desired, be interpreted as a continuation of the same curve. The optical radiation is usually considered to be synchrotron radiation not so much because of the spectral shape but as a result of the polarization pattern. The x-ray spectrum is shown in detail in Fig. 2. One can get an excellent fit to the data with a power law, and this fit

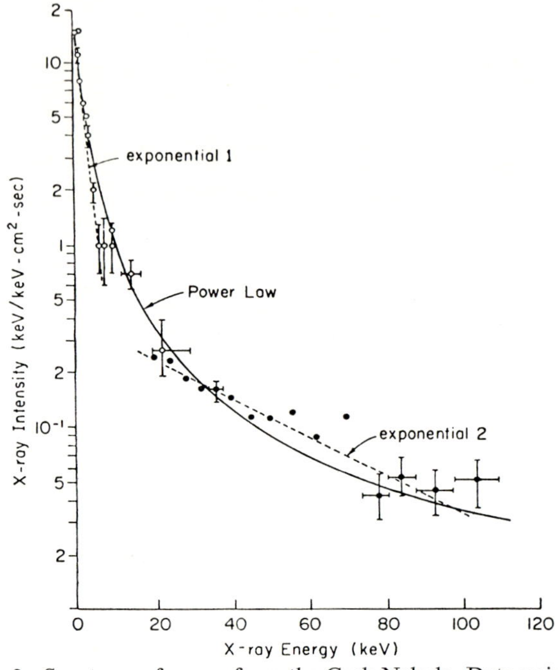

Fig. 2 Spectrum of x-rays from the Crab Nebula. Data points are from Grader *et al.*, 1966 (open circles) and from Peterson *et al.*, 1966 (solid circles). Error bars have been suppressed for some of the points of Peterson *et al.* The solid curve is a power-law fit to the data, $I(E) \sim E^{-1.5}$, and the dotted curve is a fit with two exponentials: $I_1(E) \sim \exp(-E/2 \text{ kev})$ and $I_2(E) \sim \exp(-E/40 \text{ kev})$.

is the main argument for the proposal that the x-rays represent a continuation of the synchrotron emission. A synchrotron interpretation, on the other hand, requires very short lifetimes of the order of 0.1 to 10 years.

In view of these considerations, we would like to make a completely different interpretation. We will ignore the power-law curve and instead fit the data with two straight lines (which represent exponentials on a semi-log plot; see Fig. 2). Physically, we know that a simple exponential spectrum will result from free-free transitions (i.e., bremsstrahlung) in a thin, isothermal plasma. In drawing two lines we are assuming two distinct temperatures, but we could have gotten an even better fit with several straight lines. Therefore all we are implying is that the x-ray region of the Crab Nebula consists of a thin plasma which is not isothermal.

Generalizing from our experience with the Crab Nebula, we propose that there are two classes of x-ray sources, depending on the mechanism. That is, the x-rays can be produced as synchrotron emission, giving a non-thermal spectrum, or as bremsstrahlung from a thin plasma, yielding a thermal spectrum. An x-ray source of the latter type would be a diffuse object, extended in space, and would not vary rapidly in time. Among the observed x-ray sources for which identifications have been provided we find one point source which is strongly time-variable, two others which are also time-variable whose angular sizes are not known, and the Crab Nebula, an extended object with no important time variation.

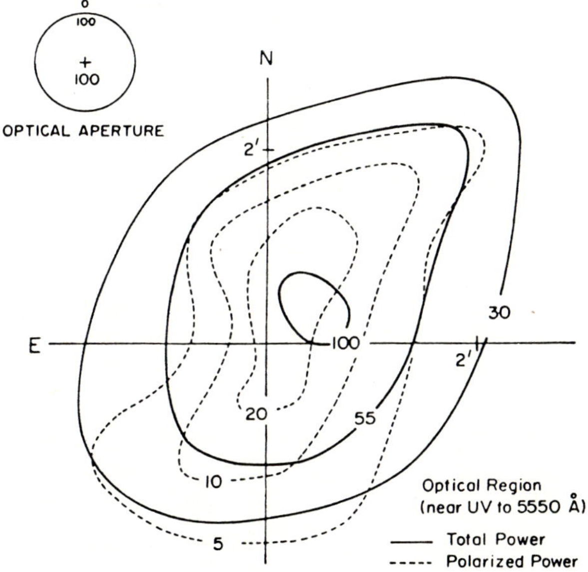

Fig. 3 Contours of total optical power and of polarized optical power from the Crab Nebula. The highest-intensity contour is normalized to 100. The aperture is drawn in one corner: full width, 1'. The band measured was the full admittance of Polaroid filters and photoelectric photometer; it includes a small emission line contribution, but is more than 90% continuum. Constructed from the data of Oort and Walraven (1956).

Now let us see whether there is any evidence that this second class of sources actually exists. We shall consider the following condition: if the x-ray source is a hot plasma containing magnetic fields, then the radio emission should be appreciably depolarized. Figure 3 shows the optical contours of the Crab Nebula. The total power in the region from 5500 Å to the near ultraviolet is

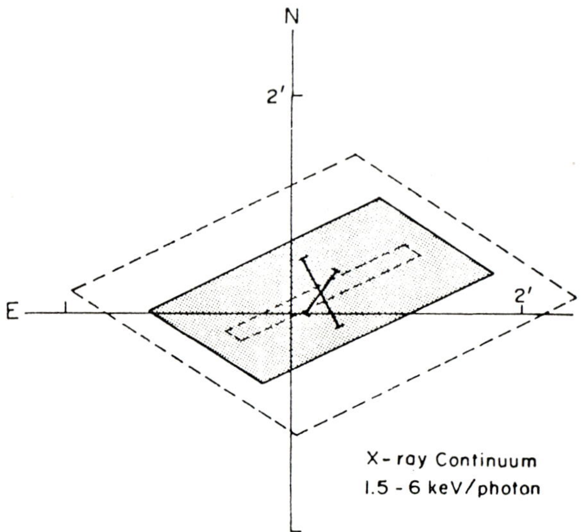

Fig. 4   The x-ray source in the Crab Nebula. The error bars indicate the quoted probable errors for the position of maximum power. The solid contour represents the most probable size for a fit to a "rectangular" shape; the dotted contours are obtained by adding and subtracting from that rectangle the quoted probable errors in size. From Oda *et al.* (1967).

compared with the polarized power in the same spectral region. There is no significant asymmetry; where the optical power is maximum there is a tendency for the polarized power to be high. Figure 4 represents the best picture we have of the object in the x-ray region from 1.5 to 6 keV. In spite of the obvious inadequacies of the data, we can conclude that the x-ray source is definitely not a point source. Furthermore, the center of the x-ray region is reasonably consistent with the center of the optical emission.

In Fig. 5 we see a radio contour map of the Crab Nebula at 15 Gc/sec. In this case the center of the total power corresponds with the center of the optical power, but the polarized power is clearly displaced from the center. This is indicative of the presence in the Crab Nebula of a strong depolarizing region which could very well be a thin, thermal plasma. A thin plasma, penetrated by just a slight magnetic field, is quite adequate to depolarize radio waves to the observed extent, without affecting the optical emission.

If we accept this plasma model for the Crab Nebula x-ray source, and take the two-line fit to the spectrum as shown in Fig. 2, we can now make some estimates of the physical conditions in the plasma. We will assume that the Crab Nebula is prolate and take the distance to be 5000 lt-yrs. The two-line fit to the spectrum implies that there are two regions of plasma at different temperatures; the respective values of kT for the two components are 2 keV and 40 keV. The electron density is of the order of 100 electrons/cm^3; the

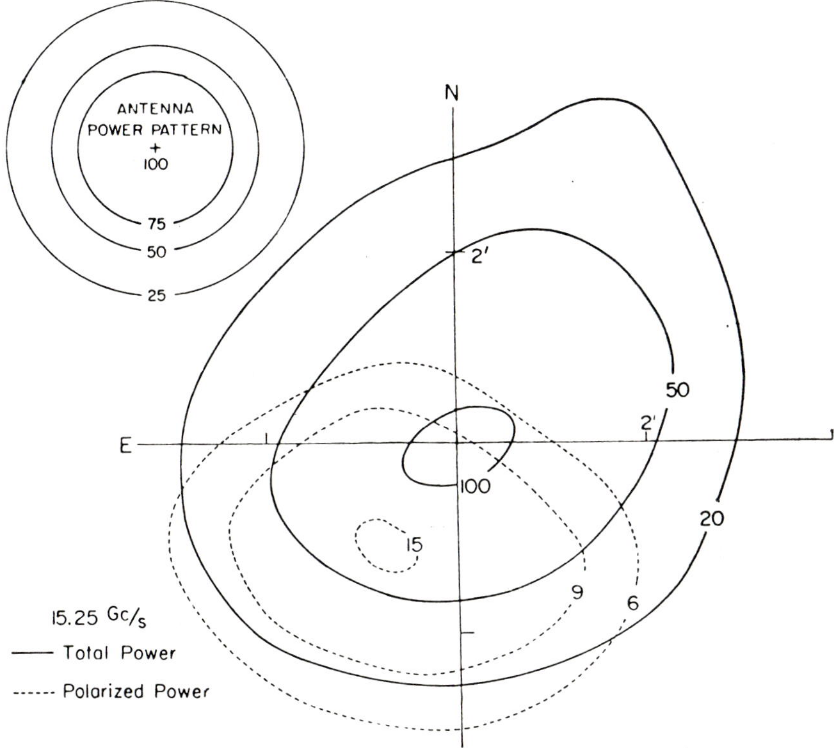

Fig. 5 Contours of total radio power and of polarized radio power. The highest-intensity contour is normalized to 100. The antenna beam power contours are shown in one corner of the figure; the full-beam width of half-maximum power is 2.2'. The measurements are made at 15.25 Gc/sec with a band width of 0.5 Gc/sec. From Allen (1967).

total thermal energy is about 5×10^{49} ergs; the mass is 1–2 solar masses. These properties will of course depend upon the distance, and the assigned values can be easily modified on the basis of the knowledge that the mass and thermal energy increase with $d^{5/2}$, while the electron density varies as $d^{-1/2}$.

Let us compare the energy in the plasma with the energies of other features of the Crab Nebula. The filamentary structure has a thermal energy of 1–2×10^{49} ergs and a mass of about 1 solar mass. The relativistic plasma which gives rise to the optical and radio emission has an energy of about 10^{47} ergs and a mass of about 10^{-4}–10^{-5} $M_\odot$. The magnetic field energy is also about 10^{47} ergs.

It seems apparent from this data that the Crab Nebula is primarily an x-ray object, energetically speaking. The optical and radio emission are secondary effects; what we are looking at is a thin, hot plasma. The Crab Nebula, by some not-understood process, is engaged in the acceleration of electrons up to relativistic speeds. The accelerating process will have a certain inefficiency, so that for every erg of energy delivered to relativistic particles, a certain amount of energy—perhaps as much as 100 ergs—will be lost as a result of the inefficiency of the process and will wind up as thermal energy in the plasma. The plasma will be expanding, but it will not lose much energy in the process because it is not really expanding against anything, so that it will radiate energy as a means of reaching equilibrium. Physical conditions in the plasma and the rate at which energy is being deposited will be such as to bring the plasma to a temperature at which it will radiate mainly x-rays.

Generalizing from this argument, we propose that any explosion which results in a process for generating relativistic particles at a time after the explosion when the material is sufficiently diffuse will deliver a majority of its energy to thermal x-rays. That is, any non-thermal radio source would be expected to have a plasma associated with it which is emitting thermal x-rays. One might also predict the existence of objects which did not produce enough relativistic particles to be radio sources, but the plasma remains and radiates as a diffuse x-ray source.

Objects which radiate as both x-ray and radio sources can be plotted on a diagram such as Fig. 6. The diagram is divided into four regions by the present levels of detectability of x-ray and radio sources. (The disparity between these levels reflects the present-day superiority of radio astronomy over x-ray astronomy.) The Crab Nebula is the only object observed as both a radio and x-ray source, although it now appears that Virgo A has been confirmed as an x-ray source and can be included in this class. Cas A, Tycho and Cygnus A are claimed, but not confirmed, x-ray sources and must remain in the tentative category for the present. In region II of the diagram we find objects which have been observed as diffuse x-rays sources and which

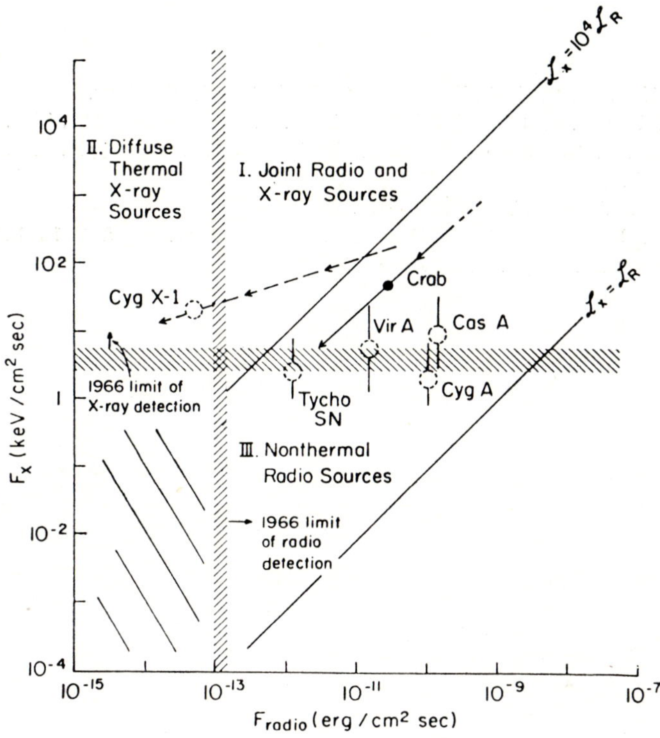

Fig. 6 A tentative set of source identifications plotted in the plane x-ray flux versus radio flux at Earth. The plane is divided into four regions by the approximate present limits of radio and x-ray detection. Region I contains sources detectable in both the radio and the x-ray band. The observed Crab source is shown as a solid circle, with an approximate time course indicated by arrows. No object except the Crab is yet known with certainty to be a joint source. In regions II and II objects would be seen as emitting either only thermal x-rays or only non-thermal radio. Dotted circles mark several possible sources; a light solid line through each mark indicates the locus defined by existing measurements. A present position for Cyg X-1 is entered in region II together with a trajectory leading from region I; this is intended to exemplify a class of object, rather than to imply a definite identification of Cyg X-1.

may be weak radio sources; Cygnus X-1 is possibly such an object. It is hoped that improvements in x-ray and radio techniques will result in the discovery of more objects which are jointly emitting non-thermal radio waves and thermal x-rays.*

References

Allen, R.J., 1967, *Ph. D. thesis*, Massachusetts Institute of Technology.
Grader, R.J., Hill, R.W., Seward, F.D., and Toor, A., 1966, *Science* **152**, 1499.
Oda, M., Bradt, H., Garmire, G., Spada, F., Sreekantan, B.V., Gursky, H., Giacconi, R., Gorenstein, P., and Waters, J.R., 1967, *Ap. J.* **148**, L5.
Oort, J.H., and Walraven, T., 1956, *B.A.N.* **12**, 285.
Peterson, L.E., Jacobson, A.S., and Pelling, R.M., 1966, *Phys. Rev. Lett.* **16**, 142.
Sartori, L., and Morrison, P., 1967, *Ap. J.* **150**, 385.

---

* This theory has been published in detail (Sartori and Morrison, 1967).

DISCUSSION

J.D.Scargle I have made some optical observations of the Crab Nebula which marginally indicate an ultraviolet excess that could be due to the optical tail of such a thermal source. I would also like to pose the question that if the x-ray source is indeed as large as the optical source, should it not depolarize the whole radio region, or at least more of it than is observed?

P.Morrison In general, this is correct; it would be much nicer if the source were completely depolarized. However, what we see may be a function of the way the field is distributed through the object.

K.Thorne Does the polarization as a function of frequency in the radio region show what you expect?

P.Morrison Yes. It had been proposed previously that the polarization as a function of radio frequency could be explained in terms of depolarization by the filaments. The same result would be obtained if we use the hot plasma instead of the filaments as the depolarizing medium.

R.Minkowski Nevertheless, one might still expect the filaments to have some depolarizing effect. Since the filaments are asymmetrically distributed, with the brighter filaments toward the northwestern end of the nebula, the shift in polarization might be a result.

P.Morrison That of course depends on whether the total effect of the filaments is large enough and whether the amount of asymmetry is significant.*

---

* *Note added May **1968**:* Downs and Thompson (Ap. J. **152**, L65, 1968) have confirmed the northwest shift of the high-frequency depolarization.

12

EVIDENCE FOR CONTINUED ACTIVITY IN THE CRAB NEBULA

Jeffrey D. Scargle

California Institute of Technology
Pasadena, California

STUDY OF PLATES taken over the past twelve years reveals that the Crab Ne
bula remains an object in which there is a considerable amount of activity
These plates were taken mostly by Baade and Münch with the 200-inch
Mt. Palomar telescope. I would like to present some of the evidence for this
activity, and briefly consider the consequences with respect to our understanding of the physics of the Crab Nebula.

Figure 1 is a plate registering essentially only the (yellow) continuum,
and shows the general structure of the synchrotron emission—to be distin-
guished from the *filaments*, which emit line radiation only. The well-known
bays on the east and west sides of the nebula should be noted. They are
strongly polarized such that the magnetic field lines, as deduced from the
synchrotron theory, curve around the contours of the bays. Hence the field
lines must extend into space at the ends of the bays, or else connect together
somehow. In either case the high-energy particles would be expected to drift
along the field lines past the ends of the bays, producing faint synchrotron
emission there (G. B. Field, private communication). Such leaked particles
would presumably find themselves in weaker magnetic fields and with lower
energies (particularly the transverse component necessary for synchrotron
radiation), so they would produce only a very faint halo around the nebula.
I have made deep 48-inch photographs with a plate-filter combination similar
to that which Arp (1964) used to study faint objects, and have found *no*
evidence of any such emission extended beyond what is found on shorter

Fig. 1 PH 1291 (Baade), 1955.7.7; 10 minute continuum (5200 Å to 6400 Å) plate without polaroid filter. (PH = taken at the prime focus of the 200-inch Palomar-Hale telescope).

Fig. 2   PH 1281 (Baade), 1955.715; 30 minute continuum exposure without polaroid filter.

exposures. This result is crudely illustrated by Figs. 1 and 2: the three-times longer exposure shows the nebula only slightly more extended.

Another feature which should be noted is the dark lane near the north-west edge of the nebula, just beyond the row of three stars. The outward motion of this feature (Oort and Walraven, 1956) can be readily followed on the sequence of plates in Fig. 4. This and other relatively slow motions (but still faster than the general expansion of the nebula at about 1000 km/sec) were discovered by Lampland (1921). The spectacular changes near the center were first noted in about 1945 by Baade (Oort and Walraven, 1956) on a series of 100-inch plates. Some of this activity is seen against the region of lower surface brightness surrounding the two stars near the center. This "hole" is part of a roughly S-shaped region approximately 80% as bright as the surrounding areas.

Let us now consider in detail this active central region, which is sketched in Fig. 3. The south-west member of the pair of stars (which are 5″ apart)

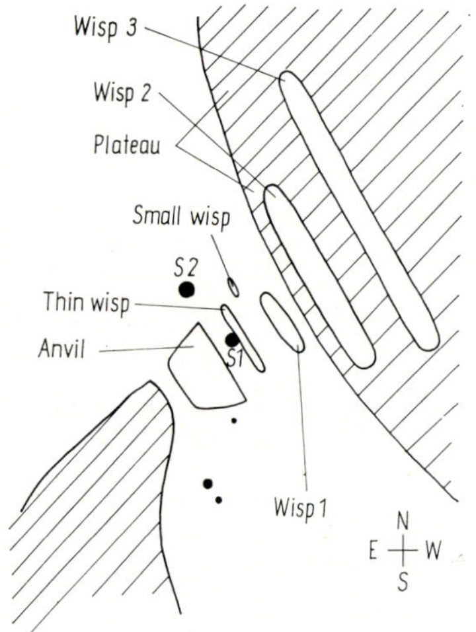

Fig. 3 Schematic picture of the central region.

near the geometrical center of the nebula has been suggested as the stellar remnant of the supernova, but at present the evidence is entirely incon-clusive. I shall call these "central stars" without implying any necessary con-nection with the nebula. The most prominent nebular feature is labeled Wisp 1, and has been referred to as "the wisp". "the moving lens", "a light ripple", etc. I shall always use the term *wisp* for this and other similar features. Almost without exception this wisp is merely the first and most conspicuous member of a *series of wisps*, more or less parallel and on the same side of the central stars. There is a clear trend for the wisps further out from the center to be longer and, somewhat less definitely, thicker. On the other hand an occasional outer wisp may appear as a small fragment, perhaps smaller even than Wisp 1. The relative inconspicuousness of the outer wisps is largely due to the fact that they are superimposed on the bright region—the Plateau in Fig. 3—in the north-west part of the nebula. It may be that the wisps are in this bright region only in projection. On some of the plates the outer wisps are quite bright, and stand out above even this bright background. The "elongated mass" in this region reported by Lampland (1921) is probably an example of a very prominent wisp far from the center. Besides the wisp series, there is almost always an extraordinary wisp not far ($\approx 1''$) from the south-west central star and approximately parallel to the other wisps. This has been dubbed the Thin Wisp because it is frequently very thin and sharply defined. On the opposite side of the central stars from these wisps is a region which also shows activity, but much less obviously. On the early plates this region was roughly trapezoidal (it is labeled the Anvil in Fig. 3) and showed little structure, but as we shall see at various times it developed wisp-like features. Finally, there are other small, faint streaks, particularly in the region between the Thin Wisp and Wisp 1. These details being very near the limit of detection of the plates, are evanescent and are mentioned only to make the point that there is undoubtedly a great deal of structure and activity which cannot be seen on even these excellent plates. It should be stressed that all of these elongated features are highly polarized. The direction of polarization is such that the wisps are elongated along the lines of the magnetic field in the nebula. This seems to be a very general relation, holding for all the structural features of the continuum radiation.

That all of the features just described undergo rapid changes can be seen by examining the time series of polarization plates in Fig. 4. These are similar to Figs. 1 and 2, with the addition of a polaroid filter oriented to transmit most strongly the light coming from the wisps. Some of the differences in these plates, of course, are due to seeing and effective exposure differences. Perhaps the most striking fact about this series of plates is that there is *always* a very definite bright wisp (Wisp 1) between 7'' and 9'' from the south-west central star. This suggests that there is a single persistent feature which moves about. Because there is no ambiguity in defining Wisp 1 it is possible to make

Fig. 4a   PH 1299 I (Baade); September 21, 1955. Note the possible wisps
in the Anvil. The Thin Wisp is rather diffuse due to poor seeing. Although
not clear on this reproduction, Wisp 2 bifurcates at its south-west end. On
the original there are at least four wisps visible in the wisp series.

Fig. 4b   PH 1315 I (Baade); October 13, 1955. Wisp 1 is definitely further from S1 than in the previous figure.

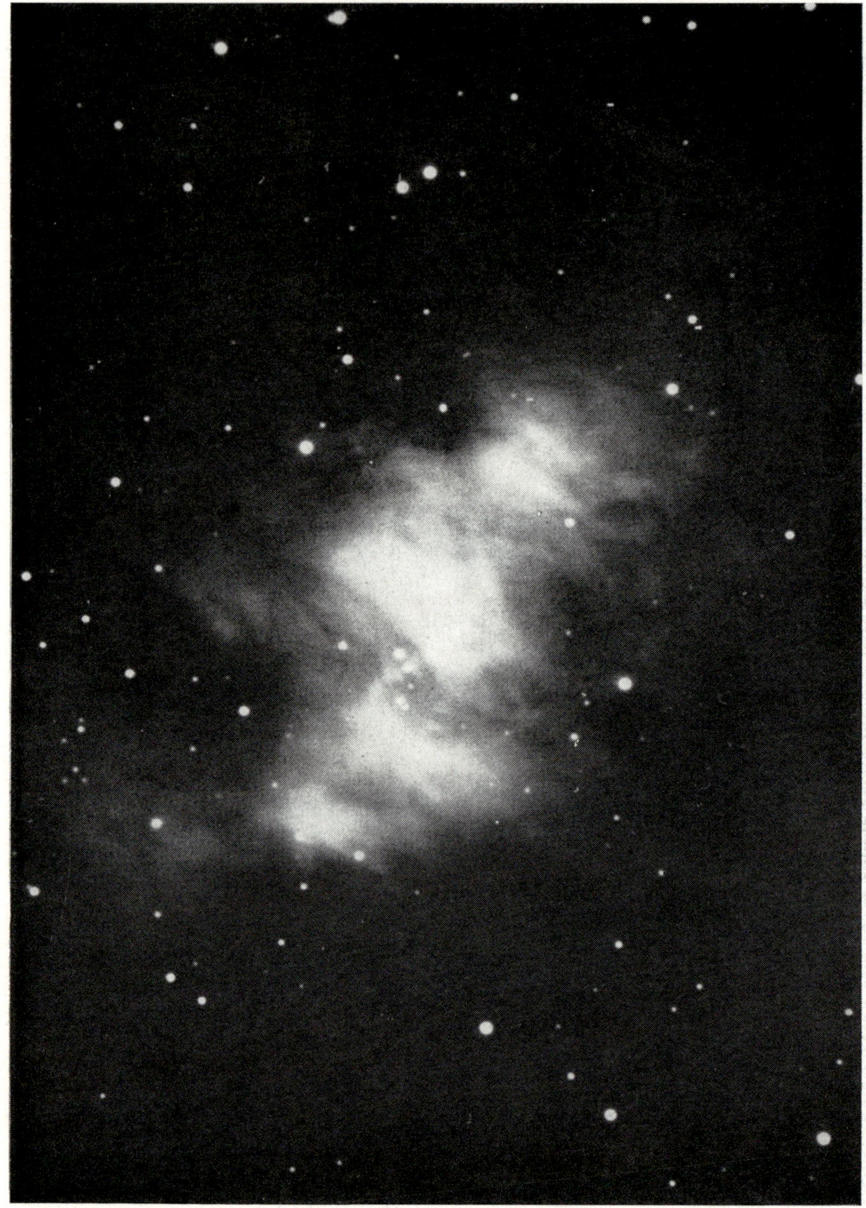

Fig. 4c PH 1347 I (Baade); December 7, 1955 (p.a. = 67.5°). Wisp 1 is at its maximum distance from S1, is over the Plateau, and hence not as distinct as usual.

Fig. 4d PH 3055 b (17 minutes); August 23, 1958. The Thin Wisp is very well defined, and there may be a small wisp between it and Wisp 1. Wisp 1 has a nearly stellar core. Note that Wisps 3 and 4 join at their south-west ends, forming a wish-bone structure.

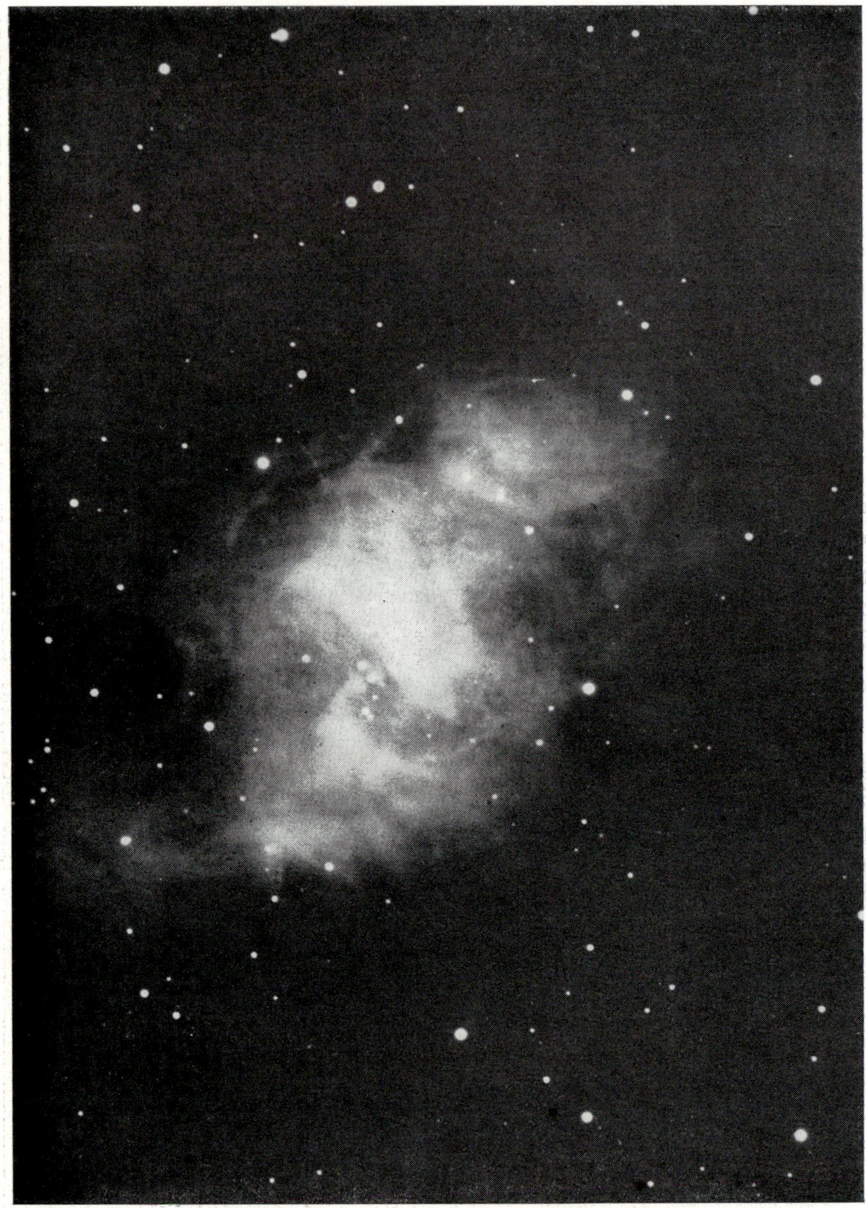

Fig. 4e PH 3442; October 9, 1959. The Thin Wisp is essentially absent. Note two wisps in the Anvil. Wisp 1 appears bent, with the south-west segment brighter.

Fig. 4f PH 3491 a; February 21, 1960. The seeing is rather poor, and only Wisps 1 and a rather broad Wisp 2 are distinct.

Fig. 4g PH 3728 b (15 minutes); December 13, 1960. The Thin Wisp is well shown and is rather far from S1. Wisp 1 has an odd structure. There are two definite wisps forming a wish-bone in the Anvil.

Fig. 4h PH 3898 b; February 3, 1962. Note the sharp, well-defined wisps
in the series. The Thin Wisp is curved and rotated from its usual orientation.
(The point-like feature is a flaw.)

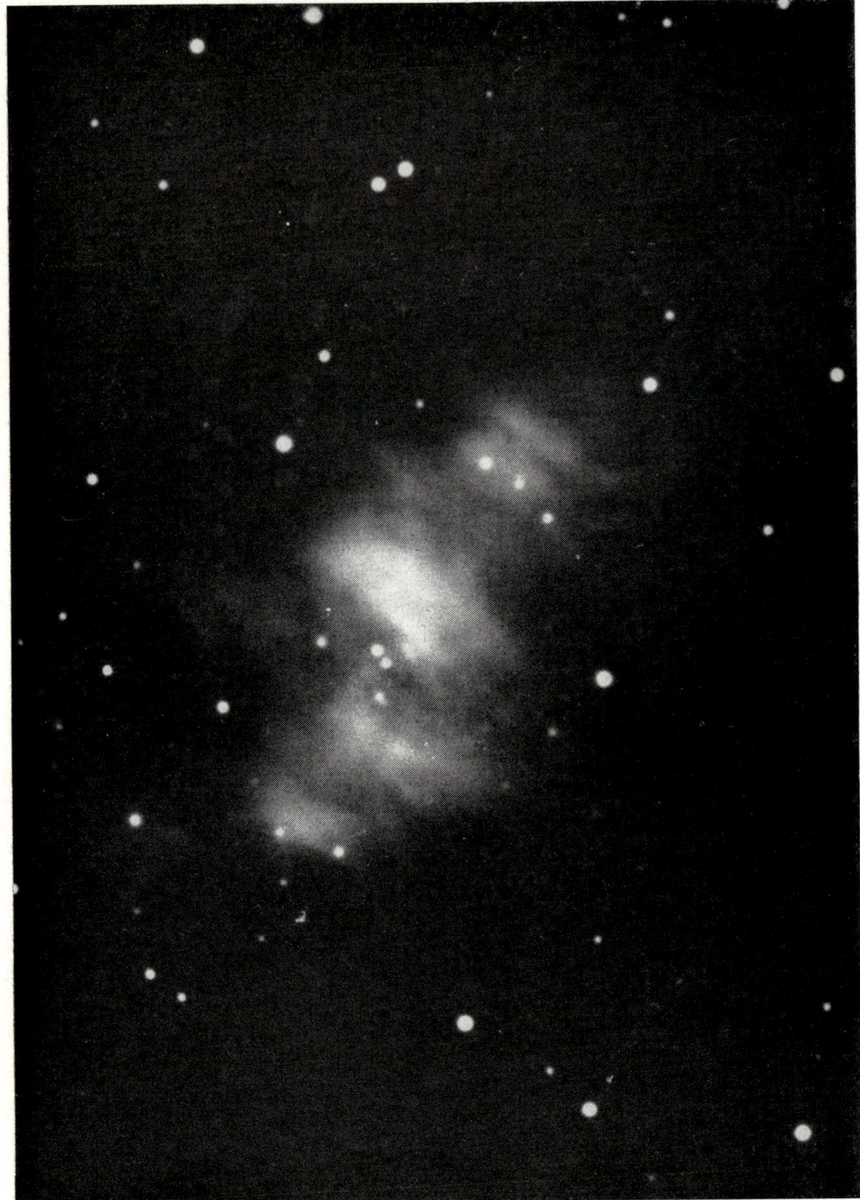

Fig. 4i   PH 4020 b (17 minutes); September 26, 1962. There is no trace of a Thin Wisp or much interesting in the Anvil. Wisp 1 is rather short. Wisps 2 and 3 intersect at their north-east ends, the reverse of the wish-bone evident in Fig. 4d.

Fig. 4j   PH 4104 B; November 30, 1962. The Thin Wisp has reappeared and is again somewhat rotated. Wisp 1 is thicker than usual, and may consist of two wisps blended together by the seeing. Note the wish-bone as in the previous figure.

Fig. 4k   PH 4321 a; October 21, 1963. The Thin Wisp seems to extend far to the north with unusual structure at this end. Wisp 1 appears to be tilted and connected to Wisp 2, and there are several more wisps extending unusually far to the north. Note the very distinct wisp in the Anvil.

Fig. 41 PH 4643 a; October 9, 1964. The Thin Wisp is faint. Wisp 1 is sharply pointed and may be composite. The Anvil shows peculiar knotted structure.

Fig. 4m PH 4986 (Arp); September 13, 1966. Wisp 1 seems to be plowing into Wisps 2 and 3, bending them back.

Fig. 4n PH 5130 b; November 10, 1967. The Thin Wisp is very bright. Wisp 2 still bends around Wisp 1 (*cf.* the previous figure), but not as tightly— indicating a kind of Alfvén wave propagation. The dark hole in the south is a flaw. Note how far the "dark lane" to the north-west has moved since the epoch of Fig. 4a.

meaningful measurements of its displacement, and the results confirm the single-wisp hypothesis (see Fig. 5). In Baade's "provisional account" of the earliest of the 100-inch plates(Oort and Walraven, 1956) it was suggested that there was a repeated creation and subsequent disappearance of a succession of wisps. In some notes to be published elsewhere, Baade had obviously realized

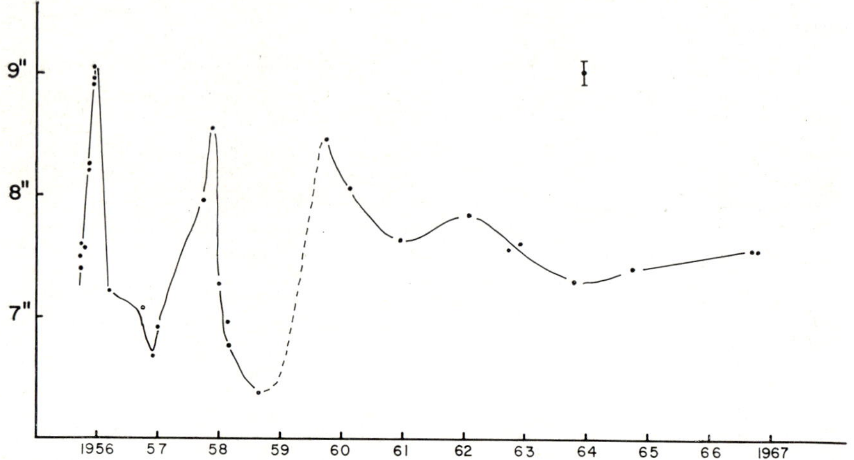

Fig. 5   Time variation of the displacement of Wisp 1, in arcseconds, from
the south-west "central" star-measured to the WNW.

that this interpretation was incorrect (based on the first few years of 200-inch plates of the Crab). Unfortunately a detailed description of the plates has not been made public until now, so this incorrect preliminary interpretation has been widely propagated. Figure 5 shows the displacement of Wisp 1 from the south-west central star measured perpendicular to the direction of elongation of the wisp itself. This is the direction of the main component of the motion, but as Baade noted there is sometimes apparently independent motion in the direction of elongation (i.e., along the magnetic field lines). From the graph it seems that in the interval from 1955 to 1960 there was quasiperiodic motion inward and outward, with an amplitude of 2″ and a period close to 2.0 years. The curve is not sinusoidal, but composed of rather sharp spikes. The slopes of these spikes correspond typically to velocities of $0.2c$ (c = speed of light) at a distance of 1500 pc. If the distance is closer to 2000 pc., as suggested by a number of authors, and the true motion is at an angle to the plane of the sky, the space velocity would be higher—perhaps

0.3c or 0.4c. While it is entirely possible that matter is moving at these velocities, it must be kept in mind that what is observed is merely an apparent motion—which could, for example, be due to a *compressional disturbance* propagating in the relativistic plasma at a velocity a good fraction of the speed of light. Such a compression naturally causes an enhancement of the synchrotron emissivity. The kinds of modes to be expected in a relativistic plasma of the sort which exists in the Crab nebula are known, and will be discussed elsewhere. To return to the observed motions, the situation after 1960 is not so clear. Taken at face value, the graph of displacements suggests a damping of the amplitude of the oscillation. On the other hand it is curious that this is just when the plates were taken less frequently and the coverage of the motion is much less complete. Thus, either there really was damping and it coincidentally started rather suddenly just when the plates were taken less frequently, or else the later plates coincidentally caught the wisp at roughly the same point in its excursions (recall that the period is very close to 2 years, and that the taking of plates is somewhat seasonal). This last idea is rather bizarre, but not an impossibility. Other changes in the wisps can be seen on the plates, and some of the more interesting events are noted in the captions to Fig. 4. Unfortunately the features other than the main wisp are generally too ambiguous from plate to plate to measure well, so their motions cannot be well-defined. There is an unmistakable trend, however, for the outer wisps to move at velocities perhaps somewhat less than those characteristic of Wisp 1. The very important point of whether the outer wisps move inward *and* outward (as does Wisp 1) or only outward (as would be expected of compressional disturbances generated by the motion of Wisp 1) cannot be decided with the presently available information. The Thin Wisp can be measured fairly easily; it is basically stationary, but shows definite motions which are of small amplitude ($\gtrsim 1''$) and probably confined to the ends of the wisp. Finally note the motion of the dark lane in the north-west quadrant. Its displacement can be readily measured, and the velocity is apparently constant at $0.7''$/year, or 0.017c at 1500 pc. Projection of this motion backward in time places the lane at the center of the nebula in 1840. Since the feature is really two bright, diffuse, wisp-like structures separated by a dark area, it is tempting to identify these as wisps which have propagated to the edge of the nebula in about 130 years.

The structure of the polarization of the Crab Nebula is illustrated in Fig. 6, which is a *composite* plate similar to that published by Zwicky (1956). Another type of composite, consisting of two identical plates taken at differ-

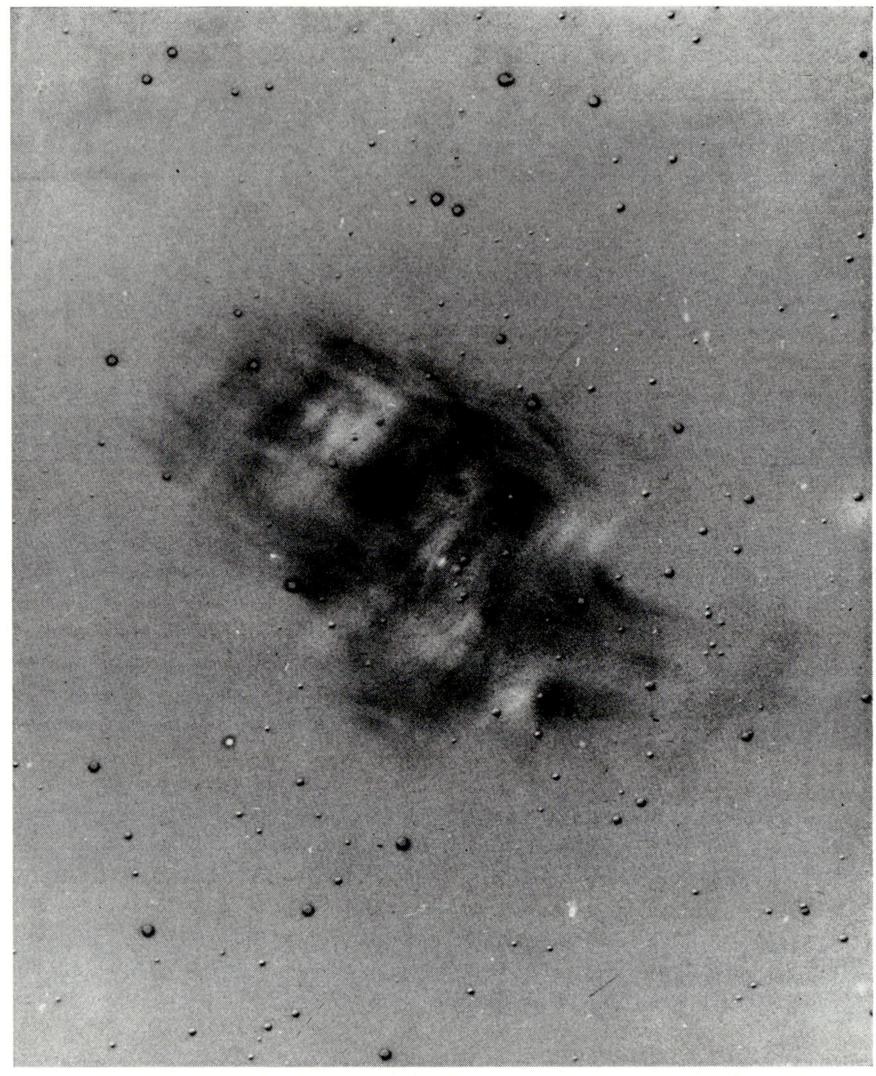

Fig. 6 Composite polarization picture, made from PH 4643a,b (position angles of the polaroid 45° and 135°), October 9, 1964. North is to the upper left, and east to the upper right. Note the polarizations of the bays, the wisps, and the "lanes" to the north-west. The oval feature in the north-central area of the nebula is a flaw.

ent epochs, can show changes in the nebula. If there were no motions or changes the image, formed by superposing a positive of one plate and a negative of the other, would be uniformly grey. The mottled structure in the north-west quadrant in Fig. 7a suggests a field of waves propagating outward (here the term "light ripples" is particularly appropriate). Figure 7b blurs these fast motions and covers too short an interval to show the general expansion; hence it exhibits the field of motions on an intermediate velocity scale ($\approx 0.5''$/year). These appear to pervade the nebula, whereas the fast motions are predominantly toward the north-west.

The effects shown on these composites could possibly be the result of photographic aberrations. Consequently microphotometer scans of the original plates were made along the lines shown on Fig. 7a. The data were digitized automatically, stored on magnetic tape, and processed on a computer. The densities were converted to intensities using the nebula itself as a calibration standard. The intensity profile for a given plate was then subtracted from the average of a number of plates, which presumably represents just the background on which the disturbances are projected. As one would expect a rather large noise signal results from this process because of granularity and a number of other effects; but a wave-like form frequently seems to be superimposed on the noise. To check this subjective result, a Fourier analysis was made of the difference profile along a length of the scan (namely $53.1''$) in the suspected active region, but avoiding the obvious wisps. For comparison purposes a similar analysis was made along a similar portion of a scan in the apparently inactive region on the south-east side of the nebula. Figure 8 shows the power spectra obtained in this way for several plates. The results for the north-west and south-east regions are indicated by solid and dotted lines, respectively. There does seem to be more power in the first region, in particular in the lower harmonic numbers (longer wavelengths). This is again subjective and not completely convincing because there is noise with an unidentifiable spectrum, and it would be difficult to predict what the random fluctuation would be in the noise spectrum. The final test, then, was to examine the phases of the Fourier components for a clear time-dependence. The phase gives essentially the displacement of the presumed waveform from some arbitrary origin. When the average phase is plotted as a function of time (Fig. 9), a very definite relation can be seen to exist. The meaning of this result is unfortunately confused by the nature of the frequency averaging, which must be carried out to reduce the noise to a tolerable level. A more complete discussion of this point is found in my thesis (Scargle,

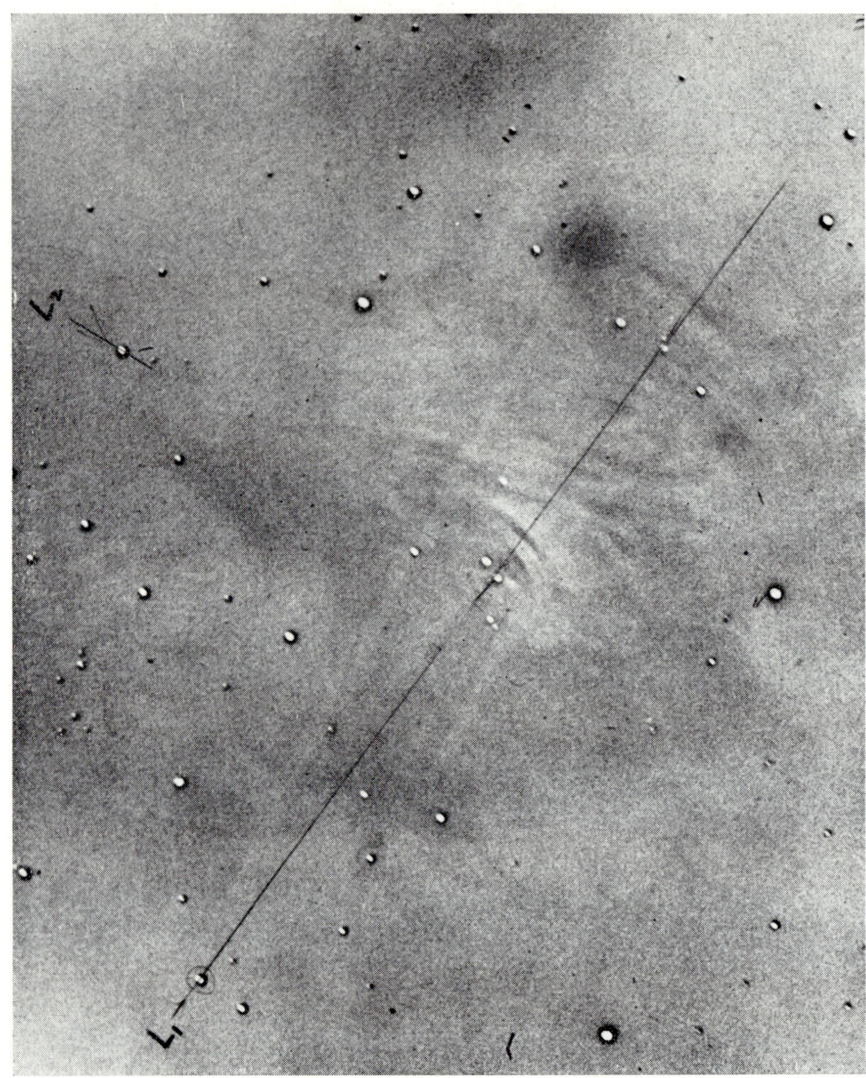

Fig. 7a

Fig. 7 Composite motion pictures: (a) 1.14 year interval (PH 3728, Dec. 13, 1960; PH 3898, Feb. 3, 1962). The cancellation is not complete, but the mottles to the north-west are probably real. The line nearly inter-

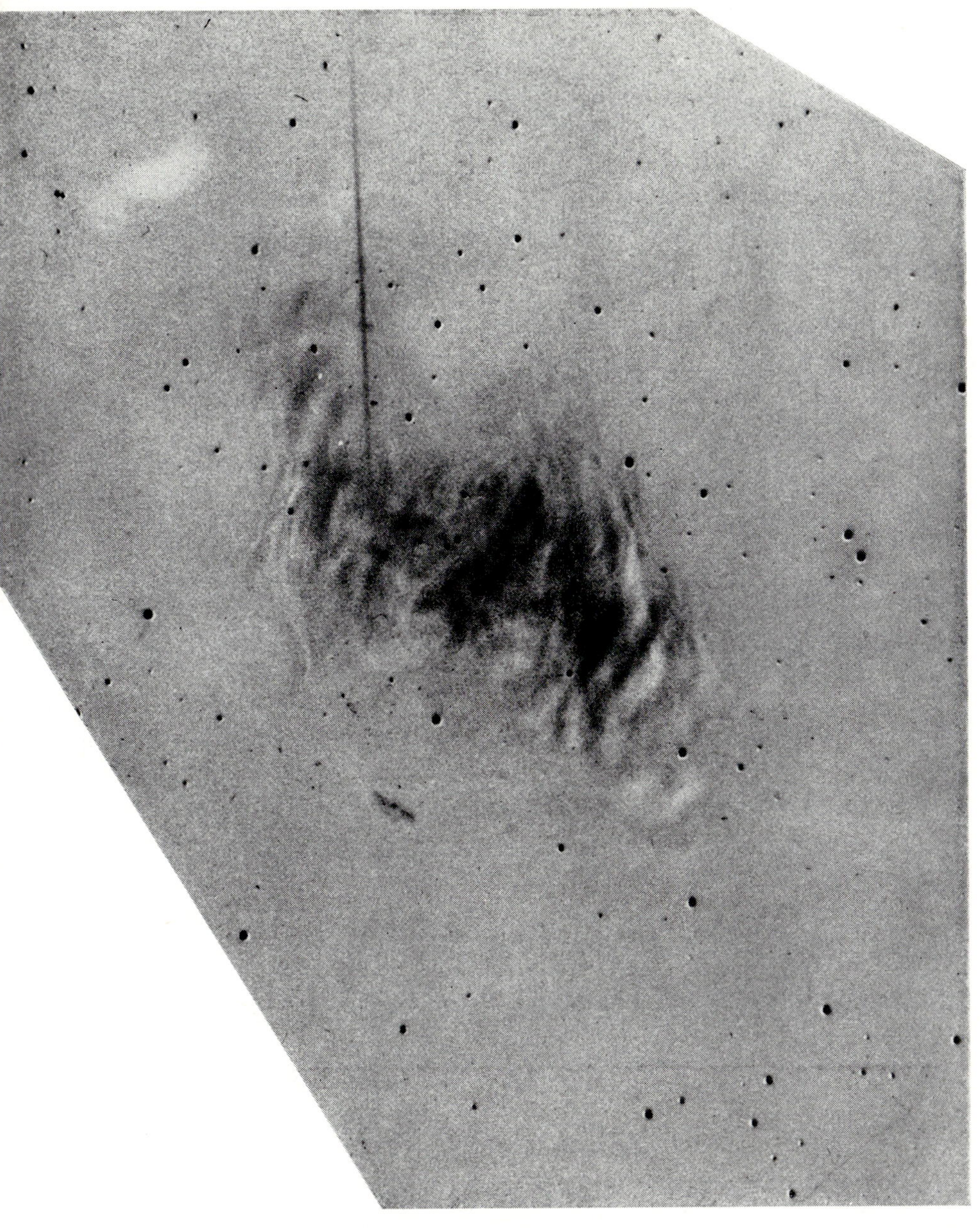

Fig. 7b

secting the central star shows the direction of the microphotometer scans discussed in the text. (b) 9.1 year interval (PH 1300, 1955.7; PH 4642, 1964.8). The general field of motions at intermediate velocities is revealed.

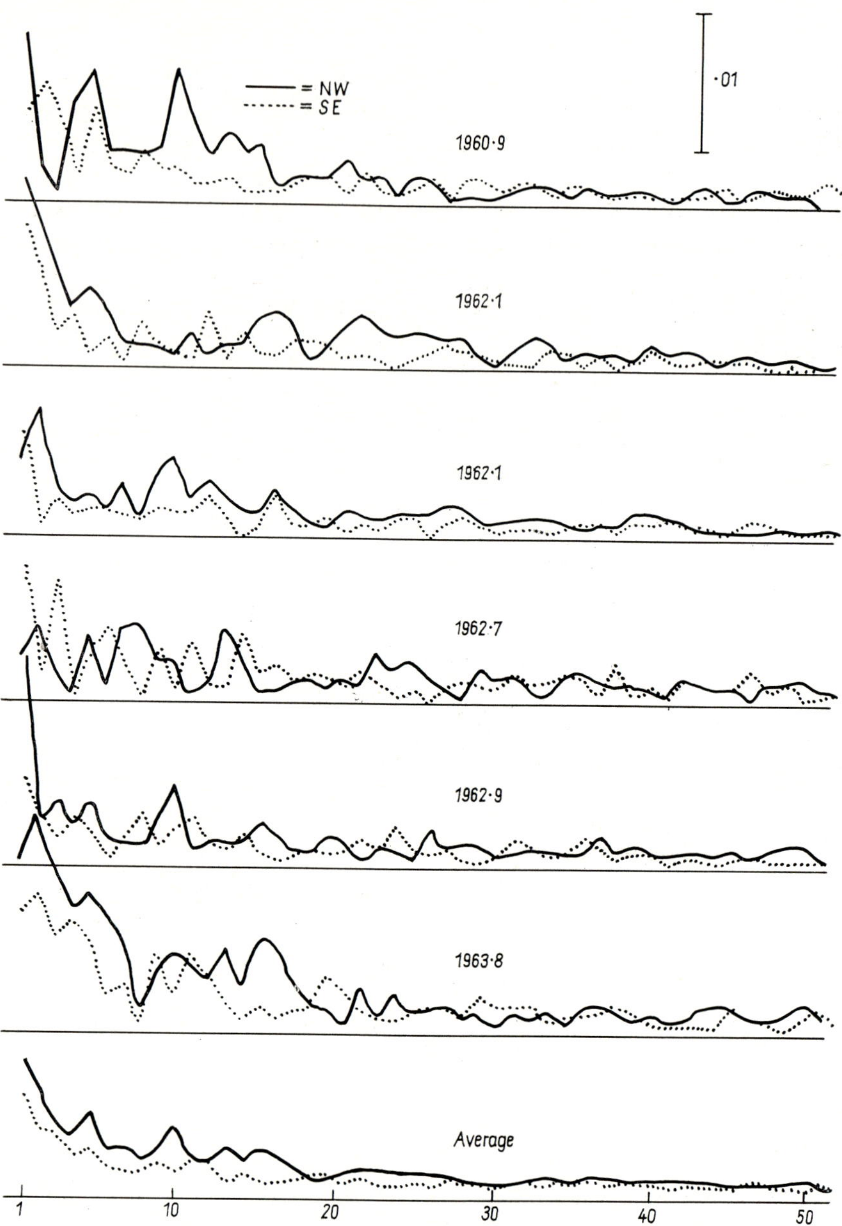

Fig. 8 Power spectra from the north-west (solid lines) and south-east (dotted lines) portions of the scan shown in Fig. 7b. Vertical scale is in intensity units such that the maximum intensity along the entire scan (i.e., where it crosses the bright core in the north-west quadrant) is 0.72 units. The horizontal scale is spatial harmonic number (wavelength $= 53''1/n$).

1968). A similar analysis for the south-east region shows a somewhat poorer correlation, but nevertheless indicates motion away from the center in that direction also. The slope of the line in Fig. 9 can be loosely interpreted as a phase velocity of $0.07c$, which is reasonable in comparison with other measurements. The period of the "waves" is about 3.5 years, which does not completely fly in the face of the conjecture that they are excited by the motion of the main wisp (remember that none of the wisps were in the region scanned). In any event, there is clear evidence for a general field of motions, presumably hydromagnetic waves (Woltjer, 1958), throughout the Crab Nebula.

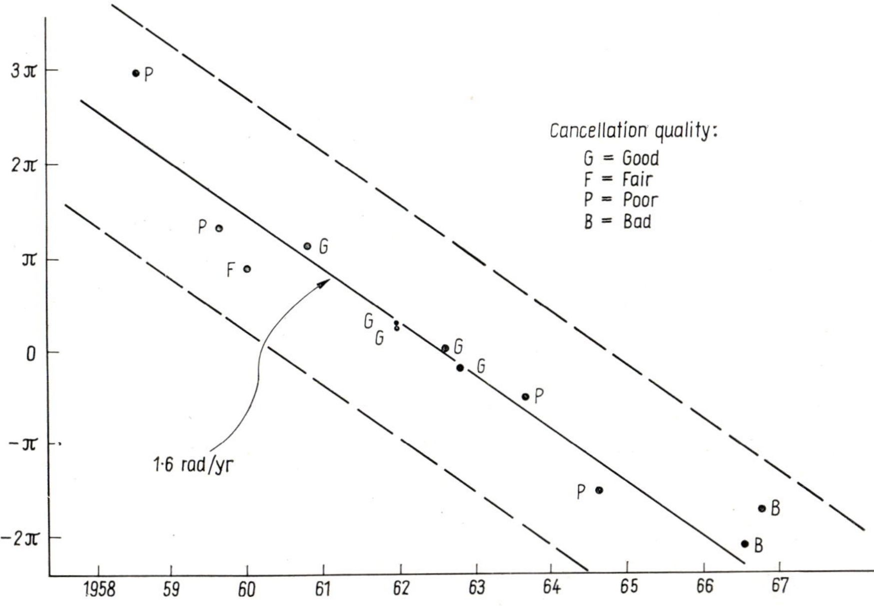

Fig. 9   Time dependence of the average phase (radians) for the north-west scan.

References

Arp, H.C., 1964, *Ap. J.*, **139**, 1378.
Lampland, C.O., 1921, *P.A.S.P.*, **33**, 79.
Oort, J.H., and Walraven, Th., 1956, *B.A.N.*, **12**, Nr. 462,285.
Scargle, J.D., 1968, thesis, California Institute of Technology.
Woltjer, L., 1958, *B.A.N.*, **14**, Nr. 483,39.
Zwicky, F., 1956, *P.A.S.P.*, **68**, 121.

DISCUSSION

G. Field   There has been some discussion recently of the core radio source in the Crab Nebula. Is there a correlation with any of these optical features?

J. D. Scargle   As far as I know the position of the point radio source is not known well enough to correlate it spatially. (*Added May, 1968:* There are gross discrepancies between the various determinations of the position of the low-frequency point source (Andrew, Branson and Wills, *Nature,* **203** (1964) 171; Kronberg, *Nature,* **212** (1966) 1557; Gower, *Nature,* **213** (1967) 1213). The last of these three observations seems to be the most free of possible error due to noise and atmospheric refraction, and is consistent with the source being located in the active wisp region. A position accurate to better than 1″ would probably allow identification with an optical feature—which I predict from symmetry considerations will be the *Thin Wisp.* This object is very close to the south-west central star, whence the required accuracy of the position.)

P. Morrison   In which direction is the active region located?

J. D. Scargle   The active quadrant is located in the north-west, toward the x-ray source. (*Added May, 1968:* The x-rays are distributed over a rather large area, so this answer is a bit misleading.)

D. Melrose   I would like to make a comment as to how a relativistic phase velocity can be achieved in a medium consisting of a thermal gas, a relativistic component and a magnetic field. One possibility is the Alfvén velocity $V_A = 2.2 \times 10^{11} Bn^{-1/2}$, where B is the magnetic field strength and n is the number density of thermal particles. Assuming $B \approx 10^{-4}$ gauss, we would require $n \approx 10^{-2}$ to 10^{-3} particles/cm³. The other possibility would involve a modified sound wave in the relativistic gas, which would propagate with a speed of $c/\sqrt{3}$. However, it will couple with a magnetic compressional wave directly perpendicular to the magnetic field, in which case the wave will propagate with a velocity $V_\varphi = (V_A^2 + V_{ST}^2 + V_{SR}^2 \varrho_R/\varrho_T)^{1/2}$ where V_A is the Alfvén velocity, V_{ST} is the sound velocity in the thermal gas, V_{SR} is the sound velocity in the relativistic gas, and ϱ_R und ϱ_T are the mass densities of the relativistic and thermal gases, respectively. For an energy density of 10^4 ev/cm³, the number density must be less than 10^{-1} to 10^{-2} particles/cm³ to give a phase velocity near 0.1 c. Consequently, although there are two possibilities for obtaining high phase velocities, both require rather low densities for the thermal plasma. The second possibility seems more attractive because the waves will propagate perpendicular to the magnetic field, and on a number of plates the polarization corresponds to a magnetic field perpendicular to the direction of motion.

J.D.Scargle *(added May, 1968)*   There are probably both Alfvén and magnetosonic modes in the Crab, although the latter are observed most readily. The low density of thermal gas required to produce the observed velocities is quite strange, but seems to be corroborated by the lack of internal Faraday depolarization. Since the velocities seem to decrease outward, the thermal density may increase outward.

13

THE CRAB NEBULA

D. B. Melrose

Belfer Graduate School of Science
Yeshiva University
New York, New York

I WOULD LIKE to turn to a number of unsolved problems relating to continuing activity in the Crab Nebula. In particular, (1) the Crab emits about 10^{38} erg sec^{-1} in synchrotron radiation and this requires an energy source operating at the present time, (2) the wisps or light ripples which Scargle has just discussed require an energetic exciting mechanism, and (3) the point source of radio emission at 30 to 80 Mc/s requires a coherent emission process which in turn requires an exciting mechanism. My chief concern is with the source of the continued energetic activity evident from these observations.

The evidence for continuing acceleration of the electrons radiating synchrotron radiation in the Crab rests on the fact that the lifetimes of the electrons is less than the age of the nebula (900 years). Most of the power radiated is in the range from near infrared to X-ray frequencies (assuming the latter to bear synchrotron). The synchrotron half-life of the optically-emitting electrons is 100 to 300 years, and for the X-ray emitters is only a few years. For the spectrum to be in a steady state, energy must be being supplied to electrons with energies from 10^{11} to 10^{14} eV at a rate of 10^{38} erg sec^{-1} to balance the synchrotron losses.

The wisps are consistent with being local compressions in the magnetic field. Such a compression from B to bB increases the volume luminosity at fixed frequency by $b^{\gamma+1}$, where γ is the index of the power law of the electron energy spectrum $N(E) \propto E^{-\gamma}$. The wisps themselves have $b \sim 5$, the actual value depending on the geometry of the compression (only the increase in surface brightness can be observed). The wisps appear at around one-tenth the radius of the nebula, i.e. a few times 10^{17} cm from the center of the nebula, propagate at about 0.1 c, appear a few times per year, and have a total energy of the order of 10^{44} erg, again depending on assumptions about their geometry. It would not be inconsistent to assume an average power input of 10^{38} erg sec^{-1} into the wisps. Throughout the remainder of the nebula less intense compressions are observed, consistent with compressions with $b - 1 \simeq 7\%$, travelling at 10^{-2} c.

Such hydromagnetic disturbances imply a central source of energy in the Crab with outward motions at velocities of at least 10^{-2} c to excite them.

The point source of radio emission is detected by interplanetary scintillation techniques (Bell and Hewish, 1967). Its angular diameter indicates a radius of a few times 10^{15} cm, situated, within observational error, at the center of the Crab. It radiates 10^{33} erg sec^{-1}, has a high effective temperature (10^{13} °K) and shows no significant circular polarization (Andrew et al., 1967). This implies that the radiation mechanism is a coherent one, and indicates that magnetic field effects are probably unimportant. The most plausible emission processes then would involve plasma waves generated through streaming instabilities.

These observations indicate that the continuing activity in the Crab originates from the center of the nebula, and one can infer that it is probably associated with the supernova remnant. One would expect an energy source generating 10^{38} erg sec^{-1} in the center of the Crab to have associated with it a thermal source radiating at least this power. However, no such thermal source is observed, the largest thermal source in the center of the Crab being a star radiating no more than 10^{34} erg sec^{-1} in the optical. Thus if there is a thermal source radiating 10^{38} erg sec^{-1} it must be at unobserved frequencies, the only plausible frequency range being the far ultraviolet or soft X-rays. I shall return to this problem later.

The transport of energy outward from a central energy source would be both radiative and through mass motions. Clearly the latter is required if the energy is to be used to excite the wisps and the point source. Such an outward transport of energy occurs following a solar flare with a large fraction of the energy in the flare going into an interplanetary blast wave.

The primary problem is to understand the nature of an energy source supplying 10^{38} erg sec^{-1}. The possible sources of energy are nuclear, gravitational, rotational, pulsational and magnetic energies. Cameron has shown us that pulsational energy cannot be stored in a neutron star for long periods of time due to the radiation of gravitational waves. Rotational energy can only be an effective energy source if there is some coupling to the surroundings or to a neighboring object, say via a magnetic field. Quantitatively, rotational energy and magnetic energy would be sufficient to account for the 10^{38} erg sec^{-1} over 3×10^{10} sec, but I ignore these here.

Nuclear and gravitational energy can act as an energy source when matter is accreted onto an object. Nuclear energy can provide 3×10^{18} erg gm^{-1} and gravitational energy up to 3×10^{20} erg gm^{-1}, corresponding to mass infall at the speed of sound in a relativistic gas. For a neutron star the surface gravitational potential energy is $c^2/30 = 3 \times 10^{19}$ erg gm^{-1}. As this seems the largest possible energy available, I assume that the energy source is due to mass infall onto a neutron star. Then to obtain 10^{38} erg sec^{-1}, mass must infall at 3×10^{18} gm sec^{-1}. Over the lifetime of the Crab this requires a mass of 5×10^{-5} M$_\odot$ accreted onto the remnant. The origin of this infalling mass could be the debris from the supernova explosion or could be from a companion object.

If the accreted matter is the debris from the supernova, then 10^{-4} M$_\odot$ must have remained with less than the escape velocity after the explosion. Colgate (1968), from velocity curves for different mass fractions ejected, estimates that 10^{-5} of the ejected mass has less than the escape velocity when the expanding gas becomes optically thin. This is inadequate, as no more than about 1 M$_\odot$ of ejected matter is observed in the Crab. Even if the remnant is a collapsed object so that mass infall supplies 3×10^{20} erg gm^{-1}, the mass required is 10^{-5} M$_\odot$. Thus only if a larger fraction than 10^{-5} is left with less than the escape velocity could this source of mass be adequate. As the Crab is anomalous in any event, and notably in having a very slow (for supernovae) expansion velocity, it may well be that an anomalously large fraction of the ejected matter is left with less than the escape velocity.

If the possibility is ignored, then the accreted material must come from a companion object. I consider two possible types of companion object. Firstly if the companion is a Jovian type planet several astronomical units distant from the primary, it is just conceivable that it would supply the necessary mass. The incident flux of radiation is sufficient to boil matter off the planet at the required rate of 3×10^{18} gm sec^{-1} provided that the radius of the planet is blown up to one third the radius of the sun, i.e. to three times the radius of Jupiter. As this model requires some stretching of the parameters, a more reasonable assumption is that the companion is a highly evolved star losing mass by evolutionary processes.

Prendergast and Burbidge (1968) have considered the transfer of mass in such close binary systems. The rate of transfer they consider is 3×10^{19} gm sec^{-1}. The model proposed by these authors has a white dwarf as the primary, their object being to account for the properties of some X-ray sources. The calculations should be little affected if the primary is a neutron star, except that the energy available is much larger, provided that this energy is not too close to the radiation stress limit (when the outward radiation pressure balances the inward force of gravity). For a primary of one solar mass the maximum luminosity allowed by the radiation stress limit is a few times 10^{38} erg sec^{-1}.

It seems desirable to avoid having to appeal to a companion object to supply the mass required. A model involving only the debris from the supernova suffers only from the difficulty of the total mass required. With such a model for the Crab, the energy supplied could be at the radiation stress limit if the remnant is somewhat less massive than the sun. I am aware of no detailed studies of models of this kind. Qualitatively one might expect the thermal velocity and the escape velocity to be comparable below unit optical depth. There would be some outward loss of mass due to free expansion and an inward motion of gas to maintain the energy supply.

Assuming that some adequate source of mass is present, the continual mass infall onto the primary will not only provide an energy source but will also lead to a freely expanding hot gas surrounding the primary. It is plausible to assume that the density of this gas falls off as an inverse square aw, i.e. if n is the number density and r the radius

$$n = n_0 r^{-2} \tag{1}$$

Presumably the energy is transported outward through this region partly in the form of mass motions. To fix the various parameters involved I assume that at $r = 3 \times 10^{17}$ cm these motions become hydromagnetic and then are identified as the wisps.

To begin with, an estimate of the thermal particle number density in the Crab as a whole is required. Assuming that the less intense variations in the Crab travel with the hydromagnetic velocity (the wisps are large amplitude phenomena and so have Mach numbers greater than unity) this velocity is 10^{-2} c. The hydromagnetic velocity is either the Alfvén velocity or the velocity of the suprathermal mode of Parker (1965). In either case this identification leads to a number density of order 3×10^{-2} cm^{-3} or less (Scargle, 1968).

Supposing that the density (Eq. 1) drops to the value 3×10^{-2} cm^{-3} at $r = 3 \times 10^{17}$ cm gives

$$n_0 = 3 \times 10^{33} \tag{1'}$$

A large amplitude shock reaching $r = 3 \times 10^{17}$ cm with a number density 0.1 cm^{-3} (the density compression is necessarily less than a factor 4) requires a velocity of 10^9 cm sec^{-1} if it is to transport 10^{38} erg sec^{-1}. With the surrounding magnetic field of 3×10^{-4} gauss (the field in the freely expanding region falls off as r^{-2} and so is necessarily negligible with respect to this value) these motions would indeed become hydromagnetic at 3×10^{17} cm, and show up as compressions in the field.

A crude estimate of the temperature of the freely expanding gas can be made by assuming that at $r = 3 \times 10^{17}$ cm ($= 1/10$th the radius of the Crab) the expansion velocity is $1/10$th the expansion velocity of the outermost regions of the Crab, i.e. is equal to 10^7 cm sec^{-1}. Identifying this with the speed of sound gives a temperature of 10^6 °K.

This model is obviously a crude one. Built into it are the requirements for an energy transport of 10^{38} erg sec^{-1} and the requirement for the excitation of the wisps. As a check one can estimate the amount of mass lost by a free expansion at 10^7 cm sec^{-1} over a spherical surface of 3×10^{17} cm with a number density of 3×10^{-2} cm^{-3}. This mass is 3×10^{17} gm sec^{-1}, in reasonable agreement with the 3×10^{18} gm sec^{-1} injected.

It seems plausible that such an energy source would have a thermal source associated with it. One can estimate the temperature of such a source by using Eqs (1) and (1') to find unit optical depth for the Compton cross section. This gives $r \sim 10^9$ cm. A black body radiating nearly 10^{38} erg sec^{-1} with this radius has a temperature of 5×10^5 °K. So a thermal source would be expected in the soft X-ray region (100–200 eV). In this range the Compton cross section underestimates the true absorption so that the temperature of the source may be lower. Even with interstellar absorption of soft X-rays (which is poorly known) one might expect to observe such a thermal source.

Another feature of the continuing activity in the Crab is the point source of radio emission. Two models for this have been presented in the literature, by Ginzburg and Ozernoi (1966) and by Zhelezniakov (1967). Neither seems satisfactory.

The model of Ginzburg and Ozernoi assumes that the emission process is Rayleigh scattering of electron plasma waves. The model requires a high plasma density, $n \simeq 10^7$ cm^{-3}, for the observed frequency to equal the plasma frequency. The temperature is around 10^6 °K and so the free-free absorption length is 10^{11} cm compared to 3×10^{15} cm for the dimensions of the source. The authors assume that the plasma is magnetically confined, presumably so that the large density drop required to allow the radiation to escape can occur. This requires a field of 0.1 gauss over 3×10^{15} cm.

However, such a field must be fixed to some centrally condensed object if it is to confine the plasma. As such a field cannot fall off less slowly than r^{-2}, this requires fields in excess of 10^8 gauss for an ordinary star ($r = 10^{11}$ cm) and 10^{18} gauss for a neutron star ($r = 10^6$ cm). These fields are impossibly large. Consequently, magnetic confinement does not seem possible; then the short absorption length appears to rule out a plasma density of 10^7 cm^{-3} and the adopted radiation mechanism must also be ruled out.

Zhelezniakov assumes the radiation mechanism to be coherent synchrotron radiation. This overcomes the difficulty with the free-free absorption length, requiring a lower plasma density. The plasma plays only a passive role in this process, relativistic electrons with a peaked energy distribution being the active radiators. This model suffers from the drawback that there is no energy source for the relativistic electrons, and the way that a peaked spectrum might be set up is ignored.

If the basic energy source for the radio emission is the plasma waves, as in the model of Ginzburg and Ozernoi, then to account for the observations the frequency radiated must be greater than the plasma frequency. One way of accomplishing this is to use the plasma waves to accelerate the electrons. This acceleration process effectively accelerates mildly relativistic electrons and will produce the peaked energy spectrum required for the coherent synchrotron process (Tsytovich, 1966). However, a further mechanism is then possible, namely bremsstrahlung generated by the relativistic electrons off the plasma waves themselves, a process closely analogous to inverse Compton radiation.

The relative importance of the synchrotron and plasma wave bremsstrahlung processes depends on the ratio of the energy density in the magnetic field to that in the plasma waves. The major difference in the two processes is the frequency radiated. An electron with Lorentz factor $\gamma = E/mc^2$ radiates synchrotron radiation at the frequency

$$\omega = 2\pi f \simeq \Omega \gamma^2 \tag{2a}$$

where $\Omega = eB/mc$ is the non-relativistic electron gyrofrequency. The plasma wave bremsstrahlung occurs at a frequency (Gailitis and Tsytovich, 1964)

$$\omega = 2\pi f \simeq kc\gamma^2 \tag{2b}$$

where k is the wave number of the plasma waves. In fact, for the latter mechanism mildly relativistic electrons can radiate at the frequency observed. For instance with a phase velocity of the plasma waves $v_\varphi \simeq 10^9$ cm sec^{-1} and a plasma frequency $\omega_0 \simeq 2\pi \times 10^5$ sec^{-1} (corresponding to $n \simeq 3 \times 10^2$ cm^{-3} for $r \simeq 3 \times 10^{15}$ in Eq. (1)), setting $k = \omega_0/v_\varphi \simeq 2\pi \times 10^{-4}$ in Eq. (2b) requires $\gamma \sim 3$ for $f = 3 \times 10^7$ c/p.

The plasma wave bremsstrahlung process requires further investigation as the radiation process for the point source is coherent (as indicated by its high effective temperature) and the conditions under which this process can be coherent have not been investigated.

A further radiation mechanism involving plasma waves, but resulting in the radiation of frequencies much higher than the plasma frequency, is that suggested by Colgate (1967). In this case a continuous scattering of electromagnetic waves off plasma waves results in an upward diffusion in their frequency. As indicated above, the requirement that the frequency radiated be much larger than the plasma frequency is imposed by the condition that the free-free absorption length exceed 10^{15} cm. The only other escape from this requirement is to ignore the observation of a finite angular size for the point source, so that the radius is much less than 10^{15} cm.

One can understand why plasma waves should be copiously generated by noting that an outwardly propagating shock passing through a decreasing density distribution can only be collisionally dominated if the collision mean free path, l, is much less than the dimensions over which the density changes appreciably. Here this requires $l \ll r$. As

$$l \simeq 3.5 \times 10^4 \; T^2 n^{-1} \text{ cm} \tag{3}$$

with the distribution of Eq. (1), $l \ll r$ requires $r \ll 10^{17}$ cm. At $r = 3 \times 10^{15}$ cm $l \simeq 10^{-1} r$. Thus significant streaming motion occurs, and one expects electrostatic stresses to be set up to compensate for the reduced collisional stresses. Under these conditions plasma waves are generated.

Thus the presence of the point source seems consistent with the overall model presented. The actual radiation mechanism however cannot yet be identified with any degree of confidence.

There are two other problems associated with the Crab Nebula which I would like to mention briefly. The first is the nature of the so-called central star. This star, if it is in the Crab, has a luminosity of about that of the sun, but is the color of an A type star. With the overall model just presented, this star could be the evolving companion, or the low frequency optical bremsstrahlung tail of the thermal source associated with the remnant. In deciding the nature of this object further spectra are required, as there remains some controversy as to whether it shows lines or not.

Secondly there remains the problem of how the highly relativistic electrons are accelerated in the Crab. Since hydromagnetic motions are observed, probably supplied with sufficient energy (10^{38} erg sec^{-1}) to act as an energy source, it is highly plausible that these lead to the acceleration. Scargle (1968) finds that there is a direction of maximum hydromagnetic activity in the Crab, and that in this direction the spectral index of the optical synchrotron radiation (and so presumably of the energy spectrum) is systematically flatter than in the remainder of the nebula.

I consider two acceleration mechanisms which might operate in the Crab. The first is physically similar to the gyrorelaxation effect (see Chandrasekhar, 1960). When a particle distribution is subjected to an impressed periodic magnetic field, pitch angle anisotropies are generated. If isotropy is re-established by some other process in a time shorter than the period of the impressed magnetic field, then the particles systematically gain energy at the expense of the impressed magnetic field perturbations.

In the Crab Nebula the hydromagnetic motions provide the magnetic field variations, and the anisotropies so generated are removed by the emission of higher frequency hydromagnetic waves (Melrose 1968a). When these secondary waves reach a state of quasiequilibrium due to their emission and reabsorption, they provide an intermediate state which allows the gyrorelaxation effect to operate.

An estimation of the acceleration resulting in this case (Melrose 1968b) shows that the mechanism is capable of accounting quantitatively for the acceleration of electrons with energies up to the energy range for the emission of optical synchrotron radiation. The mechanism can also account for the flattening of the energy spectrum observed. However, it cannot account for the acceleration of the highest electrons.

The second mechanism is analogous to a Fermi type process. This can occur when the amplitude of the wave is large enough so that the classical momentum *(e/c) A* (*A* is the vector potential in the wave) exceeds the particle momentum p. Then the changes in sign of $p - $ *(e/c) A* imply reflections of the particles from wave crests and troughs. The rate of acceleration is then just the Fermi form

$$\frac{dE}{dt} = \frac{v_0^2}{c\lambda} E \qquad (4)$$

with λ the wavelength of the waves and v_0 the velocity of wave propagation. There is the added requirement

$$\left| \frac{e}{c} A \right| > |p|$$

or equivalently, in terms of the wave energy density W_w and the Lorentz factor γ of the electrons,

$$W_w > \frac{\pi m c^2}{r_0 \lambda^2} \gamma^2 \tag{5}$$

where $r_0 = e^2/mc^2$ is the classical radius of the electron.

Applying this mechanism to acceleration by Alfven waves associated with the wisps (see Scargle 1968) one has $\lambda \simeq 3 \times 10^{16}$ cm, $v_0 \simeq 3 \times 10^9$ cm sec^{-1}, and the energy density in the waves $W_w \simeq 10^{-8}$ ergcm^{-3}. These are the most favourable values one could choose, and with these values Eq. (5) allows for acceleration up to $\gamma \simeq 10^9$, which is just the highest energy observed in the Crab. The acceleration time for the highest energy electrons follows from Eq. (4) to be a few years, about the same as the synchrotron half-life-time for $\gamma = 10^9$. Thus, with a favorable choice of the parameters, this mechanism can just account for the acceleration of the highest energy electrons in the Crab.

In conclusion it is clear that many aspects of the Crab Nebula and its continuing activity remain unexplained. The presence of an extremely energetic energy source operating in the Crab seems to require something along the lines discussed above to account for the remarkable features observed. Our further understanding of the Crab, the closest supernova remnant which can be observed in detail, will no doubt further our understanding of post-supernova conditions, even though there remains the complication factor that the Crab is an unusual supernova remnant.

This work was supported by a grant from the National Aeronautics and Space Administration.

References

Andrew, B.H., Purton, C.R., and Terzian, Y., 1967, *Nature* **215**, 493.
Bell, S.J. and Hewish, A., 1967, *Nature* **213**, 1214, and references therein.
Chandrasekhar, S., 1960, *Plasma Physics*, U. of Chicago Press, p. 58.
Colgate, S.A., 1967, *Ap. J.* **150**, 163.

Colgate, S.A., 1968, private communication. See also Colgate, S.A. and White, R.H., 1966, *Ap. J.* **143,** 626.
Gailitis, A.K. and Tsytovich, V.N., 1964, *Sov. Phys.*—J.E.T.P. **19,** 1165.
Ginzburg, V.L. and Ozernoi, L.M., 1966, *Ap. J.* **144,** 599.
Melrose, D.B., 1968a, Astrophys. and Space Sci., to be published.
– 1968b, Astrophys. and Space Sci., to be published.
Parker, E.N., 1965, *Ap. J.* **142,** 1086.
Prendergast, K.H. and Burbidge, G.R., 1968, *Ap. J.* **151,** L83.
Scargle, J.D., 1968, dissertation, California Institute of Technology.
Tsytovich, V.N., 1966, *Sov. Phys. Uspekhi* **9,** 370.
Zhelezniakov, V.V., 1967, *Sov. Astron.*—*A.J.* **11,** 33.

DISCUSSION

P. Morrison   Has anyone considered putting a star at the position of the filaments to account for the point source?

D. Melrose   That would certainly do it. An energy output of 10^{33} erg/sec is less than the solar luminosity; one would use a star as the energy source if it occurred in the filaments.

L. Sartori   If the x-rays are not synchrotron radiation, and the life-time for the optical synchrotron radiation is 300 years, I don't see why continual replenishment is still needed. The Crab Nebula would then be only three lifetimes old and if the particles have some component of motion along the field lines the lifetime could conceivably be extended to 1000 years. Isn't it then possible to trace all the energy back to the original explosion?

L. Woltjer   We must account for both the synchrotron loss and the expansion velocity, so I think that it is almost impossible to take the electrons back to the outburst.

D. Melrose   Also, we don't know what the conditions were like in the earlier stages. Presumably the magnetic field was larger, for example. However, I agree that one could explain the energy source in this manner, but it is difficult to do so.

S. Colgate   One way in which energy can be released from a magnetic field is by having flux lines which pass through a neutron star released through resistive instability. If the energy stored in a dipole field in a neutron star is released in this manner, the time scale will be of the order of the age of the Crab Nebula. A flux line released from the surface of the star by resistive instability would move out at essentially the local Alfvén speed associated with the field value at the surface. As far as the energy requirements are cony cerned, the magnetic field energy is of the order of 1 % of the binding energe of the neutron star. Actually, only 10^{-3} of the total energy is needed, sinct the binding energy is roughly 10^{52} ergs, and we must account for an outpu- of 10^{49} ergs released over 1000 years.

A. G. W. Cameron   Is it clear, if most of the internal field is wrapped up inside the star with a more-or-less quiet exterior, that field loops are going to escape, or will they decay inside and just heat the star?

S. Colgate   The problem is where does the bifurcation of the mass, i.e., that which falls in versus that which is ejected, take place relative to the field distribution? If the Crab Nebula is the result of a supernova in a star with a very large dipole field component, it is conceivable that 10^{-3} of the internal energy of the star would be released, and we would not have to consider the geometry of the field lines.

A. G. W. Cameron   Nevertheless, if we consider a region in which at one stage convection has reduced the field to a relatively small size and at other times differential rotation set in so that much of the internal field were wrapped into an internal toroid about the center, then I would think that the field would contract inward as it dissipated.

191170 WRL